INDUSTRIAL COMPETITIVENESS: COST REDUCTION

Industrial Competitiveness

Cost Reduction

by

GIDEON HALEVI

Tel Aviv, Israel

A C.I.P. Catalogue record for this book is available from the Library of Congress.

ISBN-10 1-4020-4311-2 (HB)
ISBN-13 978-1-4020-4311-6 (HB)
ISBN-10 1-4020-4350-3 (e-book)
ISBN-13 978-1-4020-4350-5 (e-book)

Published by Springer,
P.O. Box 17, 3300 AA Dordrecht, The Netherlands.

www.springer.com

Printed on acid-free paper

Printed in the Netherlands.

Contents

Preface

The objectives of industrial management are:

- Implementation of the policy adopted by the owners or the board of directors
- Optimum return on investment
- Efficient utilization of Men, Machine and Money.

In other words, industry must make profit.

Manufacturing represents only one aspect of the activities of industrial management. Present-day manufacturing methodology does not consider making profit as their primary objective.

The manufacturing process requires the knowledge of many disciplines, such as design, process planning, costing, marketing, sales, customer relations, costing, purchasing, bookkeeping, inventory control, material handling, shipping, and so on.

Each discipline considers the problem at hand from a different angle. For example, in the case of the introduction of a new product:

- Marketing will evaluate its attractiveness to the customers
- The product designer will evaluate methods of achieving product functions
- The process planner will evaluate the required resources
- Finance will evaluate the required investment
- Manpower will consider the work force demands
- The manufacturing engineer will consider floor space and material handling
- Purchasing and shipping will consider how to store the product

Each discipline optimizes its task to the best of its ability. Each manufacturing discipline has its own objectives and criteria of optimization according to its function. For example: the designer main objective is meeting product specifications; the process planner's main objective is that the items will meet drawing specifications; the production planner's main objectives are meeting the due date, and minimizing work-in-process. The profit objective is not on top of the list of any manufacturing discipline. Even if each discipline functions optimally, this does not necessarily guarantee overall optimum success with respect to management's prime objectives.

The traditional manufacturing cycle is a one-way chain of activities, where each link has a specific task to perform and the previous link is regarded as a constraint. Thus, for example, master production schedules accept the routing and bill of material as fixed (as well as quantities and delivery dates); it does not question these data and its planning must comply with them. Process planners accept the product design without question; in fact, they do not even consider the product as a whole, but rather, the processing of each item is regarded as a separate task. The capacity planner accepts the routing as given, and employs sophisticated algorithms to arrive at an optimum capacity plan.

Therefore, the chain of activities that comprises the manufacturing cycle is considered as a series of independent elements having individual probabilities of achieving a criterion. The probability of the success of any link is independent of every other link with which it is functionally associated. Thus the overall probability of the chain optimally achieving particular criteria is very low.

It is unanimously agreed that each discipline in the manufacturing process must consider the interest of other discipline interests. However, there is no practical and methodical way to accomplish this.

In this book an attempt to achieve a cooperation of all disciplines is made by organizing a meeting of all discipline managers to discuss and understand each other's problems and difficulties. Each discipline presents its task and explains the difficulties and problems the he faces. Some of the problems are due to the rigidity of the system, i.e. constraint imposed by previous discipline. A group discussion follows to validate the necessity of such constraints, and to propose a method to eliminate or ease up the constraints. The standpoint of different disciplines is considered with a view to reach understanding and acceptance of operation methods.

The book is organized in two parts. The president of a company opens the symposium defining the need to increase company profit, and to reduce manufacturing cost.

Part one deals with how to reduce inventor y cost by inventory management and control methods. Each session is devoted to a specific area, such as the objective and need of inventory;how to keep it to a mini mum; how to verify the validity of inventory records;work-in-process reduction.

Part two deals with how to reduce cost of production management. Each session is devoted to a specific area such as: product specification;product design;process planning; production planning; shop floor control.

The role of management in cost reduction is the topic of a decision support session.

Two appendices are included to further explain cost reduction methods. Appendix 1 elaborates and explains SPC - statistical process control, 6σ method. Appendix 2 gives an example of the flexible production planning method.

Gideon Halevi
12 June, 2005

GLOSSARY

C	CONTROLLER
CC	COMPUTING CENTER
D	DESIGN MANAGER
H	HUMAN RESOURCE - PERSONNEL
F	FINANCE
FM	FOREMAN
I	INVENTORY MANAGER
M	MARKETING
P	THE PRESIDENT
PM	PRODUCTION MANAGER
PP	PROCESS PLANNING
PR	PURCHASING
PS	SCHEDULING
Q	QUALITY
S	SALES

Chapter 1

SHOP FLOOR COST REDUCTION

1. INTRODUCTION

The president of the automotive part manufacturer company was not at all certain how to react to the pressure from the traditional vehicle manufacturers, and therefore jobbers and distributors to decrease prices. This was on top of the cumulative average decrease in prices of 13.3% since 1998. All the while labor, energy and raw material costs have been rising in addition to an increase in competitors in emerging markets such as China or India.

To cope with this dilemma he decided to organize a meeting with all department managers of the company to evaluate diverse proposals.

The Manager of Finance (Mr. F) raised his hand and asked to add some remarks and propose actions. He noted that in addition to the competition from the low cost countries, and the request by the customers for lower prices, there was another problem, which might turn into a bigger one. The stockholders of the company were dissatisfied with the return on their investment. In their last meeting a proposal was made to let the company operate for another six months and if the revenue had not increasd to a satisfactory level, they would recommend closing the company.

To increase the stockholders revenue our profit must be increased.

profit = selling price -/- actual cost

To increase the profit there are several methods that I propose to consider, which are:

1. Increase sales prices
2. Increase sales volume
3. Decrease employee salary
4. Decrease cost of inventory
5. Decrease processing costs
6. Call for government assistance and tax reduction

The president asked the Manager of Finance (Mr. F) to elaborate on the government possible assistance (6th method).

Mr. F explained: Companies can be competitive based on variables under their control; however, their competitiveness is directly affected by macroeconomic variables such as taxation rates and interest rates. The government should assist companies to compete better in the global industry by eliminating capital tax, lowering corporate income tax rates and lowering interest rates. Around 1.300 families' livelihoods depend on our company. If our company were to be closed they would probably become unemployed and the government would have to support them. Therefore, it is in the government's interest to assist us to survive. We can submit a request and ask for support.

The president thanked the Manager of Finance (Mr. F) for the clear definition of possible measures to take. He made the following remarks and notes for discussion.

The first option, i.e. to increase sales prices is out of the question in our present situation. We must lower sales prices while increasing profit.

I wish to remove the methods of increasing profit by decreasing labor cost. Our highly trained staff utilizes the full range of appropriate technology to ensure that each product shipped will be of the highest quality. Management policy is to keep our employees happy and be one of the higher paid companies in our field.

In principle I do not like to call on the government for assistance. We should work on our machines and not on the government.

The controller (Mr. C) noted that while considering the president's notes, there are three available cost reduction methods: decrease cost of inventory; decrease processing cost, decrease management and overhead cost.

The president agreed and proposed to start the discussion on how to reduce cost of inventory. He mentioned three modes of inventory: Raw Material, Work in Process (WIP) and Finished Products. I propose to discuss each one separately.

Let's start with raw material cost reduction.

Chapter 2

INTRODUCTION TO INVENTORY

1. THE PURPOSE OF INVENTORY

The controller (Mr. C) pointed out that "inventory" is a broad term; it includes several types of inventory for a different purpose. To reduce the cost of inventory, each type should be treated differently.

The president asked the controller to elaborate on that statement.

Mr. C explained: The competitive factors in the market for a manufacturing company are prices, and qualitative aspects including services and delivery dates. It is normal that when a customer wants to purchase a certain product he will places his order with a company that meets his required delivery dates. Therefore, delivery dates are surely a very important factor for the company to be competitive.

To meet a customer's demand, the company can take one of the following solutions:

A. Keep a very high stock of finished goods. Then, whatever demand comes there is no danger of losing the customer order, since the company can meet the demand immediately.

The drawbacks of this solution are as follows.

- Tied up capital in the finished stock can be dangerously high.
- Some of the finished stock may go to dead stock because the total demands are limited and the product life cycles are getting shorter and shorter.
- There is a need for large storage space.

- Retailers, wholesalers, and manufacturers in many cases have only limited space for keeping inventory, thus it is very difficult for them to keep all the stock.

Because of these reasons, this seems not to be the right solution.

B. Keep stock of raw material and have very short throughput times to replenish the stock of finished goods. One function of inventory is to act as a buffer between sales and production. In other words, it separates the sales function and the production function and enables each to function independently.

In its broadest perspective inventory can be defined as a matter of trying to keep the most economical amount of material in order to be able to increase the total value of profit. Inventory can also be considered in a negative sense as an asset not yet utilized: idle materials, idle machines, and idle manpower. In this sense the purpose of inventory management is to avoid having too much total idleness among an aggregate of all the assets owned.

From an investment standpoint, inventory is commercially wasteful. However, from an operating point it absorbs the difference between forecast and actual demand. Semi-finished components and subassemblies are maintained in order to:

- reduce the delivery time quoted for the end product;
- balance seasonal demand fluctuations;
- take advantage of volume discounts in purchasing and manufacturing.

Inventory control is divided into two main parts: inventory management and inventory accounting. The objective of inventory management is to keep capital investment in inventory to a minimum while maintaining a desirable service level; this is the planning and controlling aspect of inventory. The objectives of inventory accounting are to keep track of inventory transactions and to supply information required by other systems.

The use of computers in industry has made it possible to plan and control inventory as an integral part of the manufacturing system. The need for items and subassemblies is established to correspond with the exact date when assembly is scheduled to begin. These are dependent items - they depend on the master production schedule. The independent items are forecast and planned according to management policy in the master production schedule.

Conventional inventory management, with its theories of service level, economic order quantity, safety stock, and order point, was appropriate for manual systems; in spite of its unrealistic basic assumption of gradual depletion (in manufacturing, depletion tends to occur in discreet lumps because of lot sizing at higher levels), there was nothing superior. However, in the era of the integrated manufacturing system conventional inventory management has become obsolete. Its objective of keeping capital investment in inventory to a minimum while maintaining the desired service level is met more satisfactorily by master production scheduling and requirement planning. In conventional inventory management, dead stock is defined as items in stock with no issue or movement for a predetermined period (e.g., two years). A slow moving item is defined as an item with issue movement of no more than, for example, 10% of the balance within a year. Whereas these terms were suitable for the conventional system, with the computerized integrated manufacturing system the definitions should be changed. As we plan future activities we do not count on historical data; thus, for example, a better definition of dead stock would be: stock that we don't plan to use for a predetermined period in the future, where requirement planning furnishes this information. In an extreme case, if an order was cancelled dead stock might consist of items just arrived or that has not yet been received in inventory.

1.1 Inventory objective

Inventory control is central to the various manufacturing activities; in most industries the activities start and end in inventory. The received raw material is first entered into the storeroom, and then issued to the manufacturing shops, and the finished items are entered into the stockroom; items are issued for assembly, and subassemblies and finished products are entered into the storeroom; purchased components are entered into the storeroom when received; finally, shipping to customers is carried out from inventory. This procedure places inventory at the junction point of all activities, thereby making it a good source of information concerning the progress of manufacturing.

The objective of the inventory system is to supply information required by other systems; thus the inventory system is a dependent system it depends on the applications desired and on the information required by the integrated system. The inventory system should be designed according to these specifications.

The following are examples of the above-mentioned applications and retrievable information that serve as the objectives of the inventory system:

- Control over plant properties.
- Supply data about on-hand stock to the requirement planning system.
- Supply data to expediters on the availability of items required for assembly.

- Supply data for alternative materials.
- Approval of suppliers' bills.
- Supply data on the value of stock to the balance sheet.
- Supply data on material cost to the costing system.
- Control over indirect material usage.
- Supply data to estimate cost of products.
- Supply data on order delivery dates.
- Control over raw materials supplied to subcontractors (when the customer supplies the material).
- Control over quality control of suppliers.
- Supply data to suppliers' rating system.
- Supply data for calculating shop hourly rates.
- Supply data for budget preparation.
- Supply data for forecasting future sales.
- Supply data for tax considerations.
- Supply data for evaluation of different price systems in inventory.
- Supply data needed for decisions on buying or expansion of plants producing required material.
- Control over dead stock and slow-moving items.
- Control over physical count of stock.

Each of the specified objectives should be analyzed *vis-à-vis* the required data and the way that they will be handled. In addition, the reliability and data processing technique requirements should also be considered. The inventory system is constructed with these objectives in mind.

In conclusion to the elaborate inventory theory we can define three types of inventory, which are:

- Raw material(Depended Items and Independent Items)
- Work In Process (WIP)
- Finished goods

The president thanked Mr. C and he suggested that we start our discussion by considering only the reduction of the raw material inventory.

Chapter 3

RAW MATERIAL REDUCTION SESSION

1. WHY RAW MATERIAL

The Inventory Manager (Mr. I) pointed out the raw material costs about 35% of product cost. At an interest rate of about 4% that means that if we completely eliminate the raw material inventory we might save only 35% x 4% which is about 0.014 or 1.4%. This will not meet the requirement of over 3% cost reduction.

The president noted that the inventory reduction, as high as it will be, will not solve the problem of the required reduction.

2. ELIMINATE RAW MATERIAL INVENTORY ALTOGETHER

Mr. F was the first to respond and noted that he agrees with the proposal to reduce or eliminate inventory altogether. He pointed out that from an investment standpoint, inventory is commercially wasteful. Furthermore it created additional investments in order to store and manage the inventory. Warehouses must be built, workers hired to carry the goods to these warehouses, and probably a carrying cart needs to be bought for each worker.

In the warehouse people would be needed for inventory management and rust prevention. Even then, some stored goods still rust and suffer damage. Additional workers will be needed to repair the goods before removal from the warehouse for use. Once stored in the warehouse, the goods must be inventoried regularly. This requires additional workers, and probably buying computers, hardware and software, for inventory control.

All this will contribute to cost increases.

3. ELIMINATE INVENTORY ERRORS

This remark provoked the Production Management Director (PM): he rose angrily and said that from a manufacturing point of view, inventory is a must and it is not a waste. Even with the present inventory size there are problems that cause waste. If inventory quantities are not completely controlled, (which usually they are not) shortages and dead stock can arise. A single mistake in inventory management that creates a shortage of an item needed may cause the assembly line to stop, or workers being idle while waiting for the material. The waste caused by even one such mistake will eat up the profit, (saved by not keeping stock) that ordinarily amounts to only a few percent of sales and thereby endanger the business itself.

The Company Controller (Mr. C) raised his hand and asked to add some remarks and propose actions. May I draw your attention to the fact that over 3% of inventory cost might be just a figure caused by inventory system faulty data (errors). If we increase the reliability of our inventory data we might decrease cost by the required amount.

Let me explain what I mean.

Inventory accounting control (not inventory management) handles a huge amount of transactions. A moderate size company processes about 500,000 transactions every month. With such an amount of transactions, data errors are inevitable. An error rate of as low as 1% will result in 5,000 faulty data items; these errors will be compounded to an extent that depends on the number of times these faulty data items are used (some input data items might be used as often as 30 times in different applications and reports). Consequently, an input data error rate of 1% might result in an average total of 50,000 errors in data processing files and reports.

Errors can cause considerable damage, for example, possible results could include the purchase of material already in stock or failure to buy or produce an item required for assembly. Some errors produce less serious results, such as paying a debt to the wrong vendor, while others may be simply unpleasant and add to the large stock of jokes told about the stupidity of computers.

Special care is usually taken to reduce the number of errors that originate at the input end of data processing. However, the measures employed are often not sufficient, since even an error rate of 0.1% – about the lowest limit that can be expected-is intolerable. People make mistakes and there is nothing that can be done about it.

What I suggest is to examine our records and to devise a system that will reduce the number of errors and thereby reduce inventory cost.

The president and the participants were amazed by the proposal and they indicated that it was hard to believe that there could be so many errors in the inventory records, and asked what might be done to prevent them.

The Production Management Director (Mr. PM) and the Finance Manager (Mr. F) indicated that they were aware of errors in the inventory system. In many cases the records indicate that there are sufficient items for assembly or processing, but when they request an issue, the response is that the item in question is not available.

The Controller (Mr. C) indicated that there are cases where the book value of the inventory is very high. When investigating such cases it appears that the unit price is wrong. It turned out to be a unit cost where the price was for a box of 100 units. There were several cases of confusions and errors in the unit of measure and some of errors in specifying the code number of an item and the catalog books.

3.1 How to eliminate the inventory errors

The President asked Mr. C to elaborate on his proposal for eliminating inventory errors.

Mr. C explained: Inventory is a passive stage in the manufacturing cycle; it does not plan or initiate any activity, but merely serves the active stages. This fact can be used to increase the reliability beyond the general reliability measures. Moreover, it may serve as a production information and control system for companies that do not wish to control manufacturing at the operation level and are satisfied with controlling it at the part level.

Inventory transactions are not initiated by the storekeeper, but rather by one of the active stages of the manufacturing cycle. Therefore, each inventory transaction can be validated by comparing it to the planned activities. Fig. 3-1 shows the inventory file as a nucleus with many reference files as satellites. These reference files contain all planned inventory activities. Each transaction is marked by a transaction code that indicates in which reference file the initiation of this transaction is recorded. Before updating the inventory file, a validation check will be made against the appropriate file. If the transaction is found to be valid, updating will take place; if not, the transaction will be marked as an error.

The numbers on the connecting lines in Fig. 3-1 indicate the transaction codes. For example, an inventory transaction with code 01 results from a purchasing order. The transaction indicates the item code number, the quantity, the supplier, the order number, and so on; furthermore, it must contain the key to the purchasing orders file. A validation check is made to ensure that the details on the transaction are correct. This is done by retrieving the appropriate record from the purchasing orders file. If all details match, the transaction updates the purchasing record with the quantity received, retrieves the unit price from the purchasing orders file and records it, and then the inventory file is updated. Receipt from the production floor will be validated similarly by comparison with the records in the shop open order file. Receipt from other company stores will be compared with the issues from the same stores, while issues to customers will be validated against the customer orders file.

Issues to assembly will be validated against the shop assembly order file and against the bill of material file to check if the issued item is required for the said assembly and if the quantity issued corresponds with the items per assembly.

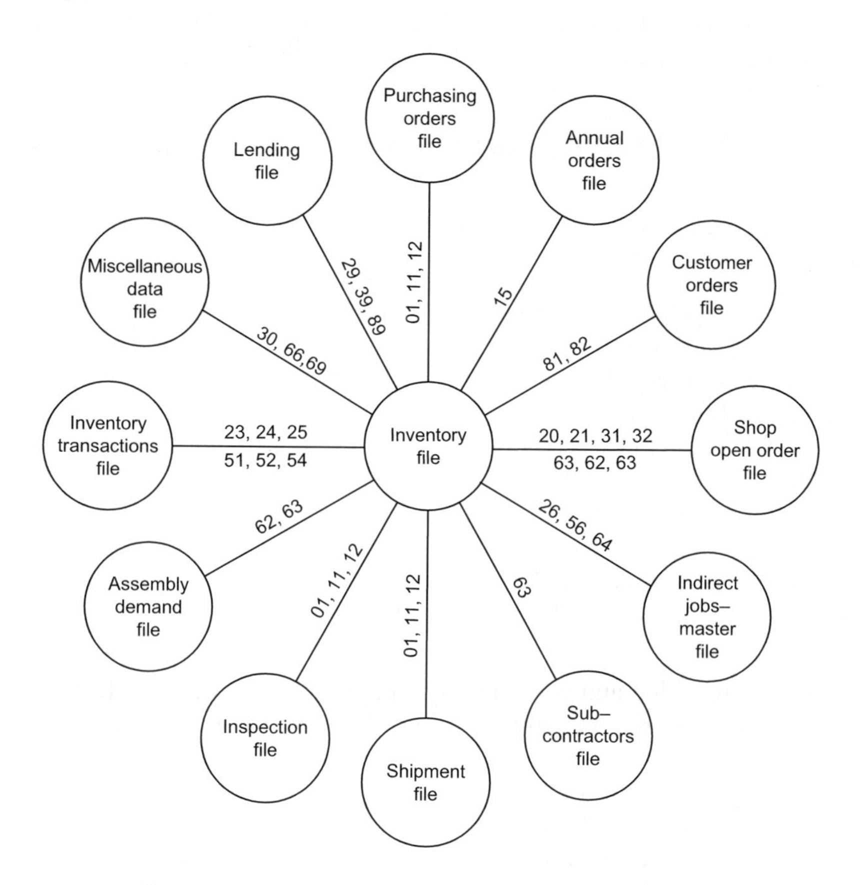

Figure 3-1. The inventory files as a nucleus with satellites

The principle of two-way data processing is applied; this saves reporting and thereby increases the reliability of the reference files. Although the reference files are used for validation checks, at the same time they can be updated if the transaction is found valid. The validation checks that the transaction was initiated by some phase, but at the same time it also checks the presence of the item in stock and of the reported quantity. Negative on-hand balance, for instance, is unrealistic.

This technique calls for retrieving all records involved from the appropriate files and bringing them into memory. It performs trial updating in the memory of all records while checking for validity. Valid transactions will update record files. The use of the data base technique is very helpful from the programming standpoint.

Not all transactions can be so closely controlled. There are some unplanned activities for which no trace and backing can be found in any reference files (e.g., issues for overhead or miscellaneous use or receipts of items purchased and paid for in cash by plant personnel). The transaction code indicates this type of transaction and no validation against a reference file is carried out; however, some logical testing can be done. For example, the value of items received must be low and within company procedure. The issues can be checked according to item type and the department that made the request.

Other types of unplanned transactions include receipt of scrapped quantity, issue of quantity to replace scrapped items in assembly, and receipt of items due to production interruptions. These types of transactions are valid and should be controlled by the production phases, not by inventory. Inventory should serve production and not control its operation. Flexibility is therefore recommended in regard to quantities. Although the incidence of transaction errors is minimized through the validation tests, inventory discrepancies still occur. Some of the reasons for this are:

- Errors in the count made in receiving or issuing.
- Errors in recording unplanned transactions.
- Entering a transaction twice or failure to enter it.
- Entering a transaction with a wrong classification code.

To establish confidence in the system, these errors must be corrected. They can be revealed and corrected by a physical inventory count. Inventory counts are a legal requirement in some places and company regulation in others.

The president was impressed by the proposal and asked for comments.

Mr. PM noted that theoretically it was a good proposal, but it was not practical. In practice it seemed that it probably would interfere with production. In many cases production needs items from inventory for supplementing rejected items, broken items, and out of tolerance items. Such cases are not planned, and can happen in the middle of production. Replacement items or missing auxiliary material will stop production. There is no time to stop the line just to follow procedures. The proposed procedures are too restrictive. Procedures should be flexible and be practical in real life. Furthermore, the need for the issue of such items is unpredictable, therefore there is no way to verify the validity of that issue. The system must allow for such issue. One may suggest solving the problem by assigning a transaction code such cases. It is a good solution, but practitioners might use it, and do use it, generally in order to bypass procedures. By using this transaction code to retrieve items for assembly, instead of by following the procedure of indicating the assembly work order, by checking if the items are according to the product structure, the quantity of each item, etc., it will devastate the production control system.

The president interrupted and asked if he did not exaggerate by saying "devastate" instead of "harming" or another more mild word?

Mr. PM responded by explaining that he chose the right word. I will explain. The manufacturing activity planning systems such as MRP or ERP computes the scheduling and purchasing by relying on data from several data bases.

The logic and mathematics upon which it is based are very simple. The gross requirement of the end product for a specified delivery is given by the master production schedule. This requirement is compared against on-hand and on-order quantities and then offset by the lead time to generate information as to when assembly should be started. All items or subassemblies (lower-level items) required for the end product assembly should be available on that date, in the required quantity. Thus, the above computation establishes the gross requirement for the lower-level items. The same computation is repeated level by level throughout the entire product structure, the net requirement of a level serving as the gross requirement for the lower level. Fig. 3-2 shows an example of these computations.

The demand for product A is specified in the gross requirement row of the product A table. There are 40 units of product A in on-hand inventory, and there is an open order to assemble 40 units, which are scheduled for period 3. The demand for 20 units of product A in period 1 will be met from inventory. This will reduce the on-hand quantity to 20 units. The demand for 10 units of product A in period 2 will also be met from inventory, thus reducing the on-hand quantity to 10. An additional 40 units will be received in period 3, thus increasing the on-hand quantity to 50. The demand for 30 units in period 4 can again be met from inventory, reducing the on-hand quantity to 20. The demand for 30 units in period 5 will be partly covered by the 20 on-hand units, leaving a net requirement of 10 units. The demand for

30 units in period 6 is not covered; this results in an additional net requirement of 30 units. Since the lead time for assembling product A is two periods, the assembly of 10 units should start in period 3 and of 30 units in period 4.

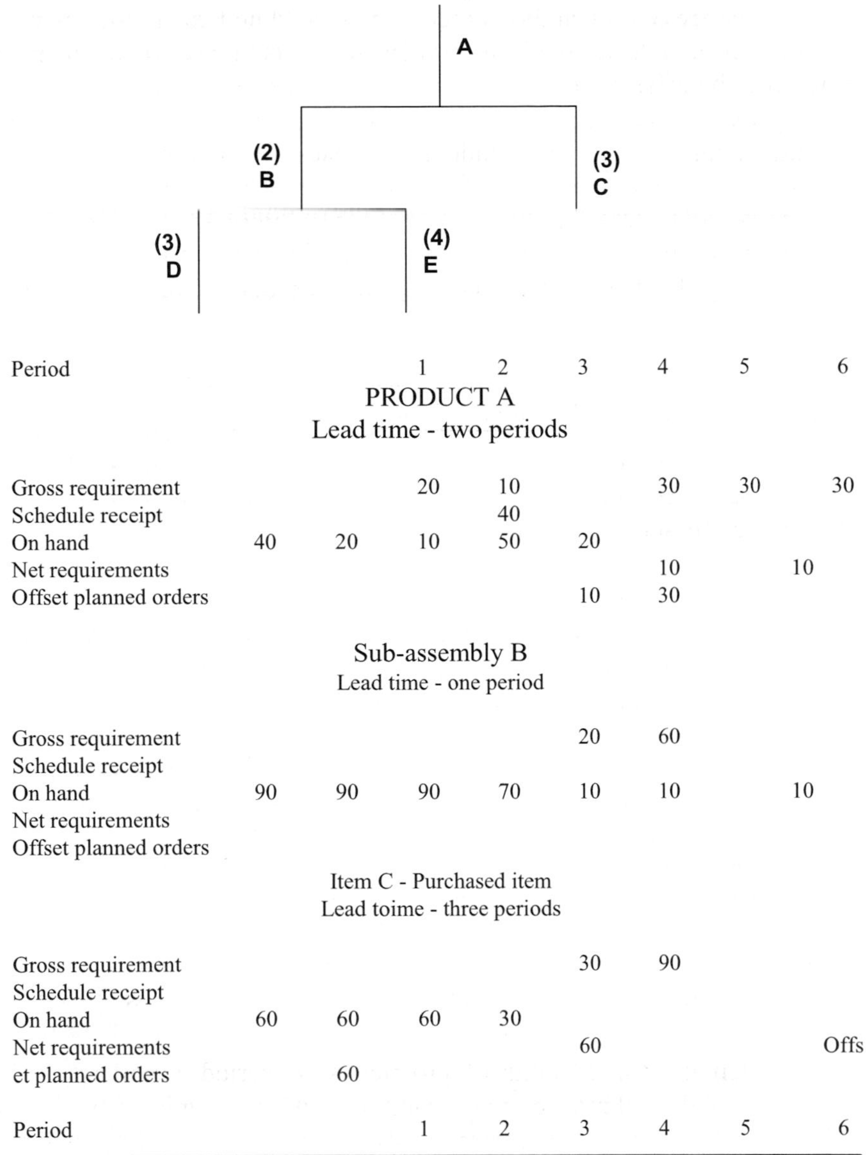

Period			1	2	3	4	5	6
PRODUCT A Lead time - two periods								
Gross requirement			20	10		30	30	30
Schedule receipt				40				
On hand	40	20	10	50	20			
Net requirements						10		10
Offset planned orders					10	30		
Sub-assembly B Lead time - one period								
Gross requirement					20	60		
Schedule receipt								
On hand	90	90	90	70	10	10		10
Net requirements								
Offset planned orders								
Item C - Purchased item Lead toime - three periods								
Gross requirement					30	90		
Schedule receipt								
On hand	60	60	60	30				
Net requirements					60			Offs
et planned orders		60						
Period			1	2	3	4	5	6

Figure 3-2. Requirement planning computations

Product A is composed of two units of subassembly B and three units of item C. Thus there is a gross requirement of 20 units of subassembly B in period 3 and of 60 units in period 4, while for item C it is 30 units in period 3 and 90 units in period 4. There are 90 units of item B on hand, which cover the demand. Thus there is no demand for items D and E. However, the 60 units of item C that are on hand, while totally covering the demand in period 3, will only partly cover that of period 4. This results in a net requirement of 60 items in period 4. Since the lead time for item C is three periods, the planned order must be offset to period 1.

Thus the activities required to meet demand are:

1. Issue of a purchase order for 60 units of item C in period 1.
2. Issue of an assembly order for 10 units of product A in period 3.
3. Issue of an assembly order for 30 units of product A in period 4.

As one can see, the logic and mathematics amount to the simple equation

Net requirement =gross requirement -/- on-hand inventory -/- on-order units.

In spite of the fact that the logic and mathematics behind requirement planning are very simple, this phase of the manufacturing cycle is very difficult to implement. Fig. 3-3 shows the relationship between requirement planning and other applications.

The implementation calls for accuracy, synchronized data from many applications and require discipline in reporting.

By retrieving items from stock by the special transaction code instead of specifying the work order for which they are intended, the gross requirement remains as calculated while the on-hand inventory is reduced (the items have already been issued). Thus the Net requirement indicates a larger amount than necessary. The result will be either a request for purchasing or issuing a production order for items that are not required, thus increase waste.

Mr. I (inventory) noted that theoretically it is a good proposal, but it is not practical. To cover all inventory transactions it will take hundreds of transaction codes. The storekeeper will need to have a PhD degree in order to function well, which probably he would not have.

Let me explain. The storekeeper is responsible for making sure that the physical count of the inventory will equal the one on the books. To be responsible he needs tools. For example, when a purchased order arrives, the storekeeper has to update the inventory records, and thus follows the purchasing records data. Recording receipt increases the item balance on books. However, the storekeeper who is responsible for the correct physical balance will insist that he has to count the items, and not to rely on the shipment records. He will probably insist on quality control inspection of the items in the shipment. This will cause a delay in the records of inventory and purchasing. The items have arrived, but neither inventory nor purchasing records are updated.

The net requirement indicates that a new order has to be issued.

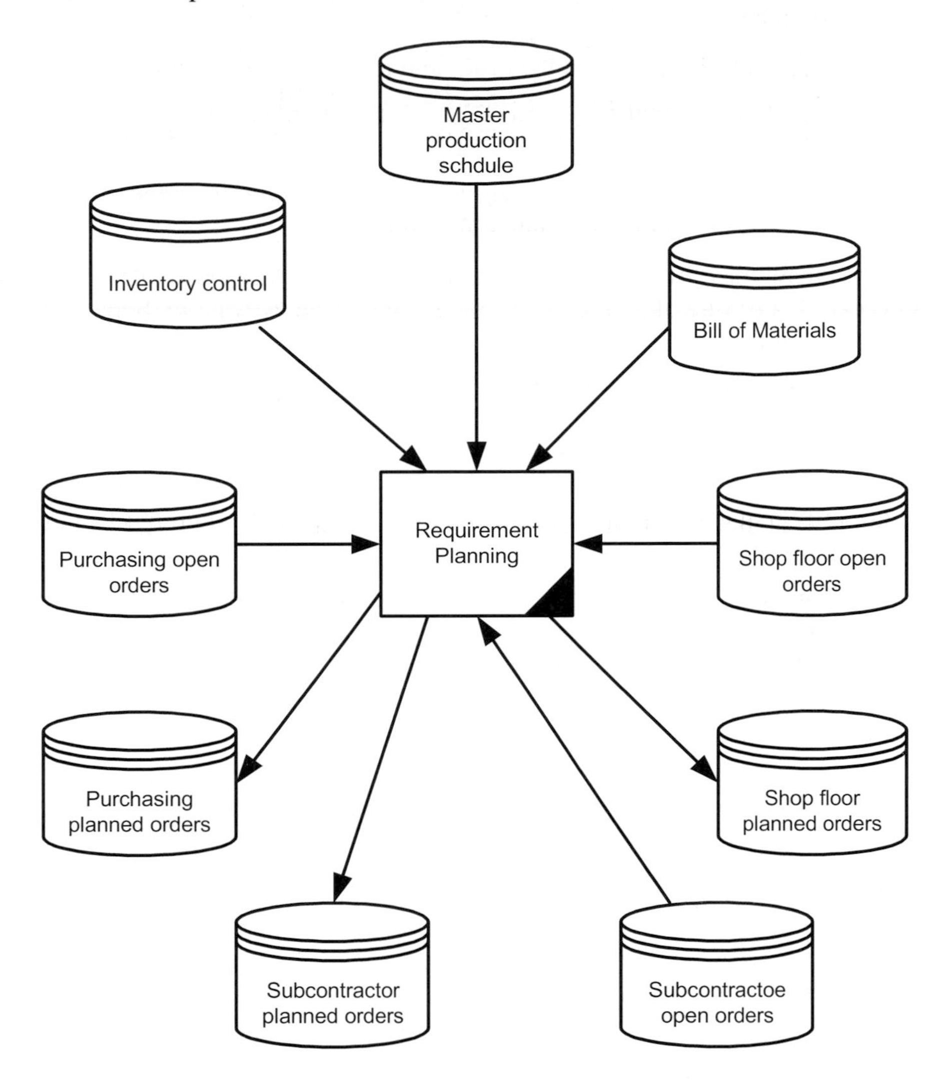

Figure 3-3. The relationship between requirement planning and other applications

In the case that the physical count does not match the shipment, record inventory will record the quantity as counted. But purchasing has a problem of how to keep its records matching reality.

The above cases can be worked out by assigning transaction coded for any possible event. For example:

Code	01	acceptable items
	05	items waiting for inspection
	06	rejected items waiting for repair
	07	return items from repair

If at that instant manufacturing activity planning will be performed, the situation will be:

Gross requirements – are as the order and product structure indicate
On-hand inventory will not consider the receipt of the shipment
On-order will still keep the purchase order open thus the net requirement

Net requirement = gross requirement -/- on-hand inventory -/- on-order units.

A system may be worked out that will cover ALL possible events; however, I am sure that the practitioners will find a way to bypass the perfect system.

I know that one may say that it is possible to enforce procedures, thus my comments are fictitious and there are no problems.

I call this a LOGICAL BUG, which is defined as:

The statement "it may be enforced" is logical, and probably the one who mentions it believes in it, but I know that it will not work, although I cannot argue against it.

Mr. F noted that theoretically it is a good proposal, but it is not practical. In practice it seems that it may "devastate" costing system.

The costing system retrieves its data from inventory transactions, shop floor job recording, from subcontractors' billing etc. Each of the above transaction relates the issue to a specific order. In the case where inventory items are issued using the special transaction code, their cost will be added to the overhead account, and not to the specific order. Astonishing cases might emerge such as producing items without raw material, or hinting that the product structure is not reliable.

Computing Center Manager (Mr. CC) remarked that from the software standpoint such a tight system can be indeed developed. Data collection hardware will be needed to increase the reliability and the synchronization of data from all disciplines. In addition the proposed system will need a smart databank software and large disk space to store the data.

3.2 Eliminate errors by physical cycle count

The President was impressed by the proposal and the arguments made by many of the managers. He recommended considering the objections.

Mr. PM said that the problem of errors in inventory records is a critical one, which causes much waste in production. Therefore he proposed to adapt the physical count method in order to overcome the errors. The proposed system as noted is a good one but not practical. It attempts to solve several problems in one system, the costing, auditing and the controlling managing. My egocentric interest is that inventory records, which are used for material requirement planning, and for resource capacity planning, will accurately present the physical amount in stock. It might be that by serving my objectives it will also contribute to other disciplines, by pointing to corrections that must be made in inventory records, assuming that the physical count is correct and not the records. The reasons for the discrepancy may be studied by the controller.

I would like to propose to employ the cycle counting method. Let me explain the method:

Traditionally, many companies have shut down manufacturing facilities in order to freeze all stock movements and take a count. Although it seems a perfect tool, errors still occur frequently. Counting is done manually, and even trained personnel make counting errors. Research on this topic indicates that when one counts more than a few hundred units of a single item, it is very probable that mistakes are being made.

Because of the inaccuracy and cost of shutting down production activities, it is recommended that "cycle counting" be used. Cycle counting is a rotating physical count at random intervals. The interval differs for each item. Some of the factors that will determine when to count are:

- Stock level. Count when the stock level is low. With fewer pieces to count, the job is easier, faster, and more accurate.
- Number of transaction activities. A dynamic item with many transactions is liable to be in error. Any item should be counted, for example, after 10 transactions.
- Item value. High-value items should be counted more frequently than low-value items. For example, high-value items should be counted every month, while low-value items can be counted only once a year.
- Physical zero balance. 'Count' any item with physical zero balance. Storekeepers know when the physical balance is zero; they should record it on the appropriate transaction, and the balance will be compared to the recorded balance.
- History of discrepancies. Items that were found to be in discrepancy in the past should be counted more frequently.

These factors, and others, should be formalized into an algorithm and be part of the inventory updating program. Thus the system can decide when it is reasonable to physically count each item. The system can then prepare a list for counting.

Reporting physical count discrepancies is done as an inventory transaction with the appropriate 'transaction code'. Such transactions will be forwarded to management for investigation.

Mr. I said that physical cycle counting can be easily implemented, it would contribute to the accuracy or inventory data file. It does not prevent recording innocent or malicious errors. But if the controller and management received misfit case reports, and study the causes, he would support the proposed system.

4. STANDARD AND AUXILIARY MATERIAL

Mr. F pointed out that previously we discussed raw material inventory, which is the major cost value part.

However, there are in inventory, also working tools, spare parts, and various auxiliary items. The spare parts are the most important, which might count for about 5-10% of the total inventory value. This inventory in many cases becomes dead stock and has to be marked out of the books. Let's eliminate this stock.

Mr. PM noted that from his point of view he would be glad if the whole lot of spare parts will become dead stock. It would just mean that the maintenance of the resources was excellent, and the resources were always available for production. You must realize that repair and maintenance personnel must be available, and it is a blessing if they can be idle most of the time, that means the processing resources are operating all the time.

Mr. C interrupted and said that he agreed with the statements made by Mr. PM. However, there are ways to reduce the cost of the stored spare parts, and the salary of the maintenance personnel.

Mr. PM explained that with the purchasing of a resource the supplier furnishes a list of recommended spare parts. The list is based on their experience with many customers who record the breakdowns, the time between events, the repair time etc. They determine the spare parts recommended list, by statistical methods that will cause the least damage to production. It is a simple equation: should we pay for spare parts, or for idle machines and loss of production.

Mr. CC said that it was all true, but statistics set a reliability factor, and the amount and items recommended is a function of the risk taking. This means if one wished to have a situation in which in 99.72% (3σ) of the cases of breakdowns the company will have the spare parts available in stock, or it may be satisfied with 68.26% (1σ). The risk is of the company and not of the supplier; therefore the recommended spare part list must be reviewed critically.

Mr. F: I do understand the need for repair and maintenance of the resources. But why should we keep the spare parts and not the supplier? Can we have a repair and maintenance contract with the supplier, that he will keep the parts, and the personnel? We have to keep such resources for our own use, while the supplier inventory may serve many users.

Mr. PM responded that he does not care who keeps the spare parts and on whose payroll the maintenance personnel is, all he cares for is that when a resource breaks down, the down time will be minimized.

The President said that it was a management decision, and he would ask the finance and economic department to look into these alternatives and recommend the best deal for the company.

Mr. F proposed to move to the next items. The cash value of working tools and various auxiliary items is around 2-5% of the total value of inventory, and usually it does not draw the attention when discussing cost reduction. However, as the president pointed out that cost reduction probably will not be by reduction in one area, but rather taking small percentages from many areas, this may be one of them.

Mr. PM raised his voice and said: do you mean that we should work with our hands, without tools, should we drill holes by hand, or clean the machines manually? Do you know what items we are talking about?

Mr. C responded: do not get me wrong, naturally you need tools to do the work, but the question is how much.

Mr. I joined the discussion and presented a few figures of inventory items:

There are 85 types of screw drivers
There are 117 types of brushes
There are 79 sizes of HS drills between the sizes of 1mm to 10 mm.

It does not make sense that we really need that many items for production.

Mr. PM responded: what is the big deal, you know how much a screwdriver, or a cleaning brush costs, it is peanuts, it makes life easy for the workers, so why not?

Mr. C explained: You are right, each item does not cost much, but each needs a storage space, keeping the records, a catalog number has to be assigned to each item.
Many misfits of inventory records will be due to recording brushes on a wrong catalog record. It will take great effort to trace and amend the records. That will probably cost more than the value of the items themselves.

Mr. I proposed to regard such items as 'open stock', which means that one record is kept for all types, invoices will be recorded, and immediate issue will be recorded for the total quantity. The items will be stored in an open space, and any one that needs them may take what he needs and return after use, all without issue or receive transaction slips.

Mr. C. responded that he liked the idea. Waste of workers' time and inventory managing expenses can be saved. However, the idea might work well for probably 10 different types of items, but not for hundreds. Moreover, 10 different types of such auxiliary items will suffice.

The president (PR) noted that it sounded like a good and worthwhile idea. What struck him most was not the cost of the items but the expenses around it. In many cases the cost of issue of a purchase order cost more than the value of the product. This procedure should be applied to all low cost items.

It seemed that the biggest saving would be by eliminating waste of direct workers. They should work on the machines or the work station, and not wondering on shop floor to and from the stockroom. It takes time off production. Moreover, on the way to stock room they might meet friends and stop to chat, they will have to stay in line at the stockroom window; all this is a waste of processing time.

Mr. CC agreed with the proposal and note that it will simplify the data collection system. By reducing the number of stock items the burden on the inventory recording system will be relieved.

Mr. I went on pointing out that it is not the cost of the items that bother, but the 'domino effect', what I mean is that once you have 79 sizes of HS drills between the sizes of 1mm to 10 mm, they follow by about 79 sizes of screws and bolts and probably rivets, several types of collets, reamers and so on.

The question is how come that we have such a large amount of items. Some say that inventory is not planned it just grows.

It might be that it is the designers' fault. When they specify a hole size the shop must have a tool to produce it.

Mr. PM agreed with Mr. I's statement and added that it also increased waste on the shop floor. I do not see why a hole of, for example Ø5.5 mm cannot do the function of Ø 5.4 mm. I suggest setting design standards. For example holes sizes between 1 to 10 mm may be in increments of 0.5.

Mr. PR summed up the discussion by noticing that the proposals were accepted by all, and proposed to implement them.

5. EXTRA QUANTITY SIZE

Mr. C pointed out that the cost of raw material in any product is between 35 to 50%. We regard this cost as a must, without it there will not be a product. However the value of raw material in stock is about 5%-12% higher than the value of raw material that goes into the products. We were amazed to learn it from the books, and we re-checked ourselves. The figures seem to be correct. We are trying to figure out the reason for this discrepancy.

Mr. PM said that he was not surprised by those findings. The raw material cost was computed by the theoretical quantity and not by the actual one. The purchased quantity is not the theoretical one, but the actual one, in order to cover unexpected events.

Mr. CC explained the steps of computing the theoretical quantity of each item per order and the real one:

Fig. 3-4 schematically represents the product structure (tree) of product A.

The letters represent items. The number within brackets represents the number of units per assembly. The number within square brackets represents the estimated percent rejects in process.

For example in Fig. 3-4:

Product A is an assembly of one item C; 4 items D and
2 sub-assemblies B. (level 2)

Sub-assembly B is produced by joining: 4 items C; 3 items E;
1 tem F; and 2 sub-assemblies G (level 3)

Sub-assembly G is produced by joining 5 items F; 1 item C;
and 4 item D (level 4)

The computations follow the following steps:

1. Explode product A (BOM) down to its elementary item. Results are shown in table 3-1, column 1 and 2. Notice that the number of units is per product and not per assembly.
2. Multiply the units per product (column 2) by order quantity, (assume 100). Results are shown in table 1, column 3.
3. Add the estimated percentage rejects to the order quantity as was computed in step 2, and presented it in column 3.

   For example: To be able to supply 100 units of product A the estimated percent rejects of 4% is added, thus 104 products should be ordered for processing.

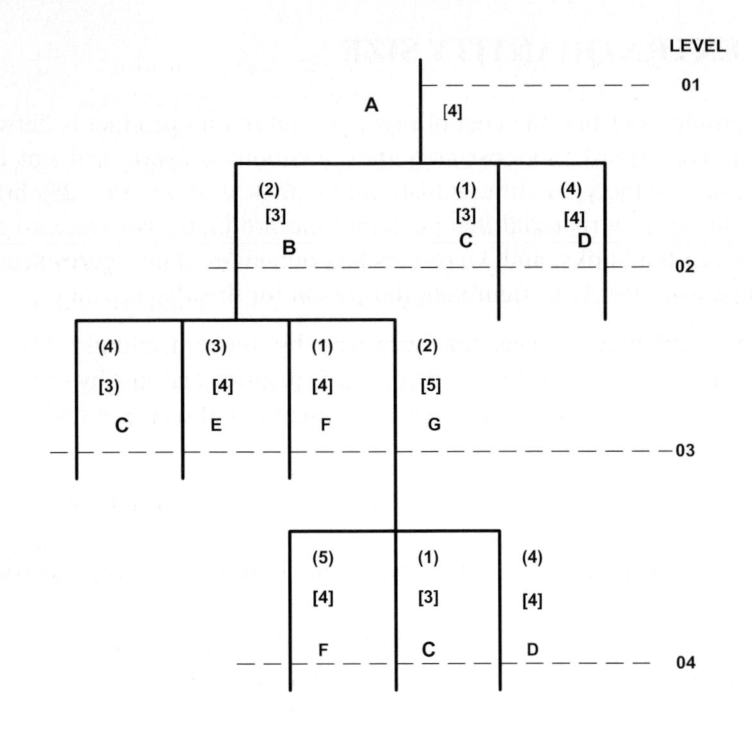

Figure 3-4. Bill of material (BOM)

Table 3-1. Bill of material explosion

ITEM 1	Units/ product 2	For 100 products 3	For 100 + ejects **4**	Total no rejects 5	Total + ejects 6
A	1	100	104	100	104
C	1	100	108	1300	1453
D	4	400	433	2000	2097
B	2	200	214	200	214
C	(2*4) 8	800	882	-	-
E	(2*3) 6	600	668	600	668
F	(2*1) 2	200	220	2200	2560
G	(2*2) 4	400	450	400	450
F	(4*5)20	2000	2340	-	-
C	(4*1) 4	400	463	-	-
D	(4*4) 16	1600	1664	-	-

2 Sub-assembly Bs are needed for assembling 1 product A. Thus for 104 units 208 Bs are needed. The estimated percent rejects is 3%. Hence 208*1.03 = 214 Bs should be ordered for processing.
4 Items C are needed to assemble 1 item B. Therefore, theoretically 800 units are needed. Considering item B rejects 214*4 = 856 items C are needed, and adding the 3% estimated percent rejects it results 856*1.03 = 882 (881.68). Results are shown in table 1, column 4.

4. For comparing the theoretical quantity to the actual required quantity columns 5 and 6 are added.

Examining columns 5 and 6 shows clearly that the hash total of the actual required quantity is higher by over 11% than the theoretical total.

Mr. PM noted that such an increase of the purchased quantity is a must. There are always rejected items in processing. Items might fall below tolerances, might get scratches, tool breaks might leave marks on the item, etc. In case that extra raw material is not available, the rejected items will be missing in assembly.

A drastic example might demonstrate such case. Suppose that the rejects of items C are as anticipated, i.e. 1453-1300=153 items. That means that there are only 1300-153=1147 items C. Processing sub-assembly G uses the required 400 units. That leaves in stock 1147-400=747 items C.

Processing sub-assembly B requires 800 units C, but only 747 are available, and 4 units are needed for one B, i.e. 747/4 = 187 (186.75) that only 187 items B can be produced. And no item C remains for the final assembly of the order, i.e. item A.

The result of not purchasing the extra raw material for item C (153 units) will result in:

Not being able to supply even a single unit of the order (item A)
And
There will be:

- 26 sub-assembly G
- 13 item F
- 39 item E
- 187 sub-assembly B
- 400 items D

left over.

That means that the purchasing of extra material saves money and it is not a waste.

Mr. CC said that they were impressed by the demonstration. It is true that extra material should be purchased, the question is how much. Your assumption was that item C will be rejected at all levels of the assembly, and not only in the processing of item C. If we assume that there are no rejects in any assembly, and we order only 1300 units of item C, (the theoretical quantity) then the results are different from the ones you presented. Ordering 1300 units with 3% rejects means that there are 1300/1.03 = 1262 good items.

400 items are used in assembling item G, which leaves 862 in stock

800 items are used in assembling item B, which leaves 62 in stock

100 items C are needed to assemble product A, which means that only 62 products can be delivered.

My assumption is as yours, i.e. it is unrealistic. There will be rejects during the assembly operation, and there must be extra stock to cover rejects. The question is how much?

Mr. C was impressed by the arguments presented. However, he noted that the estimated percentage of rejects is a statistical figure, which means that it was based on a decision of the adopted confidence level.

The president asked for clarifications.

Mr. C continued. A histogram of past reject items is made for each item. The histogram for item C will probably will have a curve, as shown in Fig. 3-5.

The X-Axis presents the quantity of the extra inventory.

The Y-Axis represents the estimated percent rejects

For example with reference to item C:

When the extra quantity was 160 items only in 0.3% of the cases items were missing for assembly.

When the extra quantity was 106 items only in 0.5% of the cases items were missing for assembly.

When the extra quantity was 90 items only in 1.3% of the cases items were missing for assembly.

When the extra quantity was 50 items in 30% of the cases items were missing for assembly.

When the extra quantity was 10 items in 90% of the cases items were missing for assembly.

Therefore the estimated number of rejects used in the product explosion is a function of the confidence level that management desire.

Reducing the confidence level from 0.3% to 0.5% may allow the reduction of inventory by as much as 33% (from 160 to 106 items).

In case that management can be satisfied with confidence level of 5% then the extra order should be for 90 items, which means a reduction of inventory by as much as 43% (from 160 to 90 items).

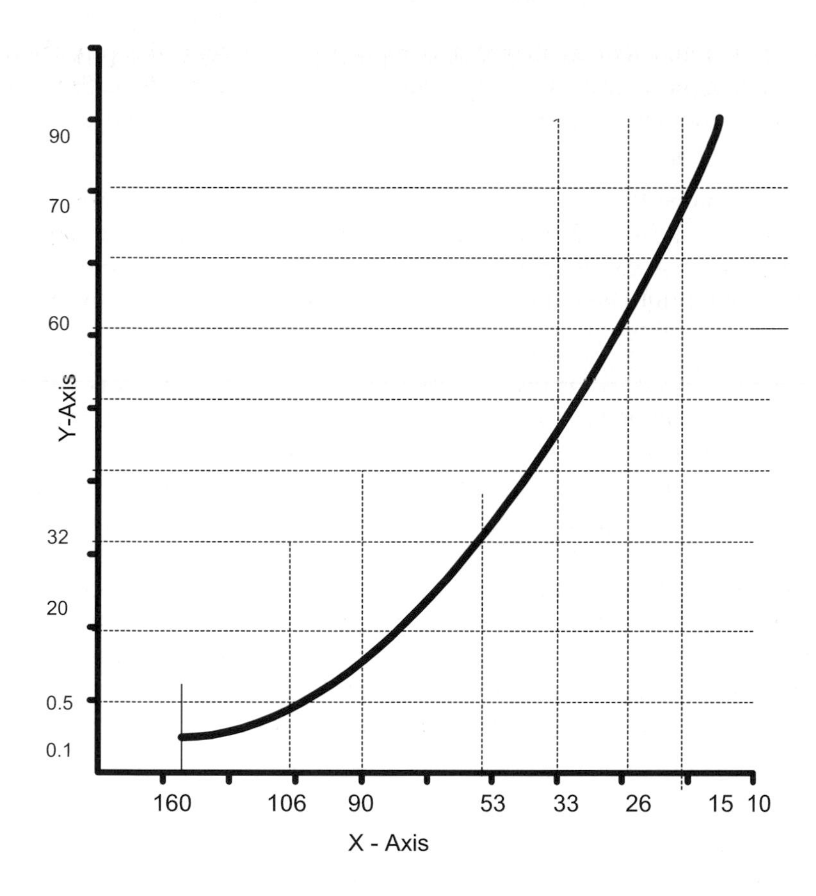

Figure 3-5. Percent of rejects as a function of extra inventory

Mr. CC noted that he accepted the presented analysis and conclusions. Number of rejects is a statistical figure, which varies from one lot to another. Rejects may be caused in processing and/or during the assembly process. The decision of how much extra to order should be based on economic considerations, the risk taking, the financial status of the company and probably the cost of the item and its criticality in the assembly of the product.

Mr. C added that the proposed logic refers to one order. However, if the same item (C) is used by more than one product, should the extra quantity be added for each item separately? It seems to me that from statistical point of view, it is not probable that in both orders the number of actual rejects will be the same. It is more probable that, if the process is under control, the left over from one order might be enough for use in the other order. What I'm saying is that in such cases the extra quantity should be further reduced. It is a complicated statistical problem.

Mr. PM noted that as it was a complicated problem, the practitioner's rule of thumb is to order the full quantity in case one product, 50% of the quantity for the second product, 25% for the third product and nothing for any other products.

Mr. F listened very carefully to the discussion and noted that the conclusions indicated that it was advisable to use standard items and materials, as the extra quantity to cover rejects can be reduced. The inventory cost might be reduced in some cases by over 80%.

Mr. Q Raised his hand to clear some of the proposals that were made. There are two sources for raw material rejects: material that was purchased by outsiders and rejects processed in our machining departments. The purchasing order indicates the rejects allowances. At receipt, the quality control department inspects the shipment to determine if it meets the order specifications. The specified reject rate in the order is the one that should be used for estimating the extra quantity. SQC (Statistic Quality Control) should check if the received items meet the required specifications. Inspection segregates the rejected items from the good ones, so there should not be any problems in assembly. The in-house produced items should be controlled by SPC (Statistic Process Control) and the control range limits. This method predicts the percent of rejects, and controls it.

Mr. C responded that this is correct, but having the natural distribution of the inspected items, it is then up to management to set the accepted rate of rejects, and thus to decide the spread, or whether to work with 3σ, 6σ or any other value that establishes the confidence level that management desire.

Mr. PM noted that the values of accepted reject from outside vendors, or the control limits for inside processing are determined by process planning and item (product) design.

The president PR interrupted and said: we are diverging from the main topic of this session which is inventory. We will have other sessions to discuss the cost effect of product design and items process planning.

6. ORDER POLICY

Mr. F noted that the purpose of the inventory system is to keep down inventory costs, and at the same time to eliminate shortage. Therefore decisions regarding order quantity and order points should be made simultaneously. Company's order policy may significantly affect inventory carrying cost. Order policy covers decisions of what to order, when to order and how much to order. Order policy does not include the topics of selecting the vendor and negotiating terms.

The president agreed that order policy should be defined, and asks the participants to come up with proposal.

6.1 Controlling order size

Mr. F explained this idea. Raw material is needed to produce the products. Processing orders do not start at the moment that the order is accepted. Therefore, instead of placing an order for raw material once a year for the quantity required, let's place a monthly order with 1/12 of the total quantity.

The size of the order has a significant impact on the average inventory level, as can be seen by Fig. 3-6. It shows that the average inventory level can be reduced from 600 units; if the yearly quantity is purchased by two orders, to 200 units if the yearly quantity is purchased by six orders.

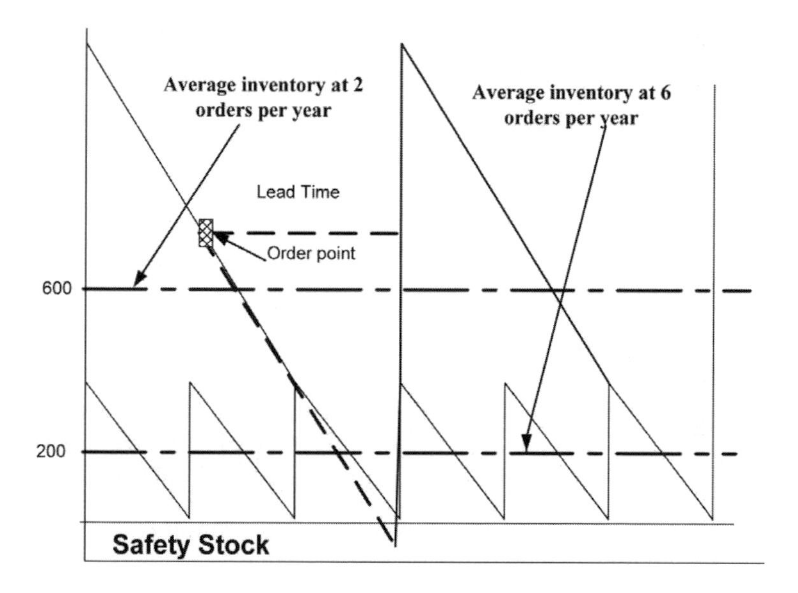

Figure 3-6. The impact of order size on the average inventory level

The benefits are in reduction of inventory carrying cost, but also in storage space, reduce damages and repairs, etc. Through control of the order size policy, management can regulate the level of inventory.

The Purchasing Department Manager (Mr. PR): I want to remind my colleagues that to issue and follow up order costs quite a remarkable amount. The saving by keeping a low inventory level must compensate these costs.

Mr. I agree that splitting an order into several batches, in addition to the extra costs in the purchasing department, increases the burden on warehouse personnel. We should check the overall economy of adopting this policy.

Mr. FM responded that he does not care about the extra work that purchasing and warehouse will have. I do care about being short of raw material, which might stop production. Splitting order quantity arbitrarily by dividing it by arbitrary numbers of batches might result in increased cost on the shop floor.

The president asks for more ideas.

6.2 Economic order quantity - EOQ

Mr. C proposed to adopt the EOQ - Economic Order Quantity as the new company policy. Instead of dividing the order into several arbitrary batches, let's do it economically and use the common EOQ method.

The president asks for elaboration on the EOQ method, and how it controls inventory level.

Mr. C continued: The theory of economical lot sizing is illustrated in Fig. 3-7. As the order quantity is increased, the average level of inventory rises, and the carrying cost therefore increases at a constant rate. On the other hand, as the order size increases, acquisition cost (e.g., set-up cost) can be spread over more items, and the unit cost therefore decreases. The total cost curve in Fig. 3-7 represents the sum of the carrying cost and order cost curve. The point of minimum cost indicates the most economical lot size.

Many techniques are used to calculate the economical order quantity (EOQ). The equation

$$EOQ = \sqrt{\frac{2 \times order\ cost \times annual \cdot usage}{inventory \cdot carrying \cdot rate \times unit \cdot cost}}$$

This can also be written in the form

$$EOQ = K \cdot \sqrt{\frac{annual \cdot usage}{unit \cdot cost}}$$

This form is easy to use and works well for items subject to a fairly steady demand; therefore, it has found wide acceptance in manufacturing.

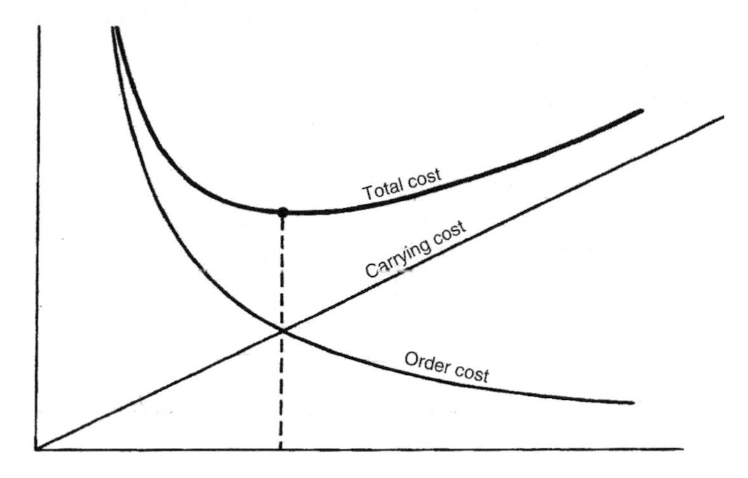

Figure 3-7. The theory of Economic Order lot size

The order quantity EOQ will be placed at the "order point" see Fig. 3-6. To assure the availability of stock, the order-point is set. Inventory is reduced gradually until it reaches the order point, at which time an order is released. Inventory continues to be depleted until the order quantity is received. The order point ensures that till the arrival of the new order there will be sufficient items to serve the needs. However, if inventory is depleted at a higher than average rate there is a chance that there will be no inventory to serve the need. Safety stock is required to absorb a higher than average rate of demand during inventory replenishment. Fig. 3-7 shows how safety stock prevents falling short of raw material.

Mr. PM observed that this technique assumes that the annual usage is known and that inventory depletion is gradual. In manufacturing, these assumptions are often not true and thus the equation ignores the timing of requirement. Therefore the standard EOQ approach is not recommended for dependent items.

Mr. CC said that the term "dependent item" is not clear, please explain.

Mr. PM explained that there are two kinds of items in stock. For example: demand for a finished product is independent of any other demand, while the demands for items that compose the product depend on the order of the product, therefore they are referred as dependent items.

For items which are required for assembled products, requirements typically are anything but uniform, and depletion anything but gradual.

Inventory depletion tends to occur in discrete “lumps” because of lot sizing at higher levels.

Components are often not available when actually needed because they have been ordered independently of the timing of each item requirements.

The situation is illustrated in Fig. 3-8.

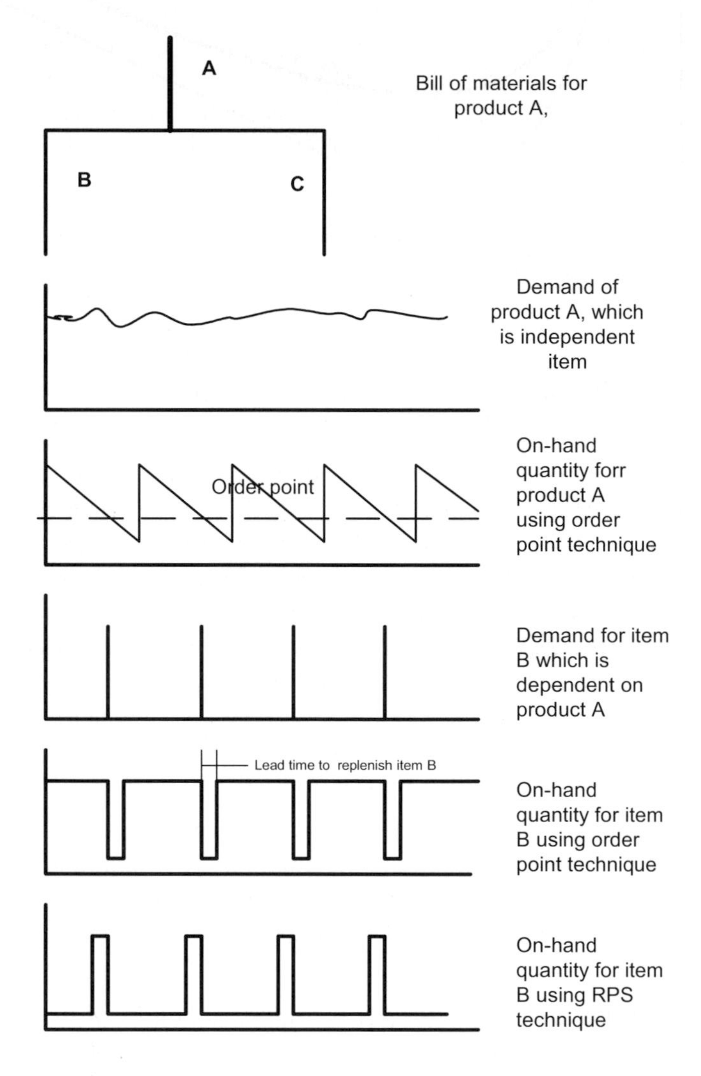

Figure 3-8. Comparison of on-hand inventory with order point technique and requirement

Only when an assembly order is placed for finished product A is demand for Item C generated. The demand for item C is very discontinuous. Maintaining a safety stock of, for example 20 units when faced with periodic “lumpy” demand of 100 units does very little good. Assuming a lead time of one period to replenish, the on-hand quantity of item C is kept unnecessarily high until the next shop order for finished product A is generated.

The ideal situation is represented in the bottom graph of Fig. 3-8. The object is to schedule the production of item C such that it arrives just before being needed in assembling finished product A. In this case, the inventory of item C is carried for only a short length of time.

Mr. CC was impressed by the requirement planning technique to reduce level of inventory. It serves only the dependent items stock, but not for the independent. However the product itself is an independent item; how is its EOQ set? It seems that this decision will affect the inventory level of all the dependent items.

Mr. I: I do not see the question; if it is an independent item then it is treated as any other item, as explained before.

Mr. PM interrupted and said that he disagreed with Mr. I. Independent items are regarded as items for general use, such as: auxiliary items, low cost items, but not products. Setting EOQ (lot size) for products belongs to the tasks of production planning and scheduling responsibility and not that of inventory.

Mr. I agrees that the product EOQ is being set by marketing and customers and master production scheduling. However, it is the lot size of the product and not of the dependent items. For those items the method of setting lot sizes by inventory control may be applied.

Mr. PM disagrees, you must remember that lot size in production has nothing to do with purchasing, using order point, safety stock. Production is in house, and is controlled by a production planning system that sets its own lot size and schedule. Scheduling dependent items objective is to serve the assembly of the product. If any item is missing for assembly then all other items wait on shop floor (WIP) and the assembly station might be idle, thus increase WIP. The importance of the master production schedule is that it is the key to the success - or failure - of detailed production planning.

6.3 Raw material for independent items

Mr. I: Order policy should specify what, when and how much to order. Conventional inventory management with its theories of service level, economic order quantity, safety stock, is based on the unrealistic assumption of gradual depletion. There is a unanimous agreement that they are not fit for dependent items inventory, but instead of abandoning this policy, they are used for independent item inventory. In my experience, depletion of independent items, in most cases, is by no means gradual. They are also dependent on the activities on shop floor. For example, packaging products for shipment, maintenance activities, requires inventory items, but at random.

Therefore I propose to use a minimum stock as order point. That means that whenever the stock of an item reaches a certain level, an order will be issued. The quantity will be determined by using rolling averages and statistics.

Mr. PR agreed that EOQ calculations do not take into account certain practical limitations that affect actual order size. Some of the more common order size limitations are minimum and maximum quantities, package or container size, storage space, joint replenishment consideration. Therefore I support Mr. I's proposal.

Mr. CC: This idea sounds very good. There are no problems to have on each inventory items, record a field stating its minimum quantity, and alerting to issue order. File history can be used to recommend the economic quantity. Furthermore, for each item and supplier, the records can predict the lead time to replenish. Thus this proposal covers: what to order, when to order and the quantity.

6.4 Raw material for dependent items

Mr. PM: Dependent items are items that are not sold or ordered by a customer as such, but are incorporated into other products. They are needed in lumps according to assembly schedule. They are dependent on independent items but on other dependent items as well, such as sub-assemblies. The chance that all items are available for assembly is very low.

Production planning and scheduling plans activities to be performed in order to meet the goals set by management. The technique is requirement planning (MRP and such). The logic and mathematics upon which MRP is based are very simple. The gross requirements of the end product for each specific delivery is compared against on-hand and on-order quantities and then offset by the lead-time to generate information as to when assembly should be started. All items or subassemblies required for the assembly should be available on that date, in the required quantity. Thus, the above computation establishes the gross requirements for the lower level items and sets the time when they will be needed. There is no sense in keeping these items in inventory when no one needs them. They should be ordered and be delivered at the scheduled time of assembly. Such a situation is represented in the bottom graph of Fig. 3-8.

Mr. FM: Fig. 3-8 seems very convincing by showing a drastic reduction in stock. However, this depends on how realistic the scheduling is. The shop floor environment is dynamic and activities are seldom performed as planned because all kinds of disruptions, and partly because the scheduling is done with infinite capacity.

Mr. F: I understand that the proposed technique schedules requirements by using a simple mathematical formula. It sets the quantity and the date. It does not consider economic order quantity, which might reduce inventory further.

Mr. PM responded that EOQ does not fit in our case. A part-period balancing (also known as least total cost) gives better results for most items that are subject to fluctuating and possible discontinuous demand. A part-period is one part held in inventory one period. Each part-period incurs a certain carrying cost. The number of parts held in inventory multiplied by the number of periods is multiplied by the carrying cost and results in the total cost. When the accumulated carrying cost exceeds the order cost, an order is planned.

6.5 Supplier contracts

Mr. F: The problem that we are trying to solve is to decide when to purchase and what quantity in order to keep it in inventory for the time that it will be needed. However, the real goal is to have all the raw material needed at the time of need. It actually does not matter how it is supplied. Keeping inventory is one costly means to meet the objective. I propose to have a contract with a supplier that he will furnish the required raw material at short notice. The benefits are savings in inventory costs, but more than that by increasing stock diversity. A system like this was proposed for spare parts (section 4).

I have experience from another job with a system where we sent out a request for a quotation to suppliers that handled a variety of products. The quotation referred to their catalog, including quality and delivery terms. Once the supplier/s was/were selected, each user of our company could call him and ask for what they needed. The system was changed from pull to push system. No inventory was kept in our shop for the items included in the contract. The benefits of this arrangement were: obtaining better unit price, eliminating the need to count and inspect shipments, eliminating most unpacking, eliminating excess material spoilage, and eliminating the stocking of inventory.

Mr. PM: It all sounds very nice, but our production efficiency depends now on an outside supplier that promised to get the required material just in time. What if he does not stand up to his promises? The idea of such a contract is good but I suggest that we keep a safety stock or a buffer for just such situations.

Mr. Q: I am concerned with the quality of the deliveries. I understand that each department may call the selected supplier and ask for delivery that will be delivered right to the user with no incoming inspection. Are we going to put our product quality and good name in the hands of an outsider?

Mr. PR: It all sounds very nice, but I am concerned that as there is no control on orders, and each department can order material right from the supplier, that instead of reducing inventory cost it will increase.

Mr. F: The proposal was in general terms and just pointing a trend. Your comments are valid and should be covered when working out the details.

Chapter 4

WORK-IN-PROCESS IN LINE MANUFACTURING

1. INTRODUCTION

The objectives of automated line manufacturing systems are to eliminate work-in-process, minimize manufacturing lead time (throughput) and quickly detect quality problems. Traditional automated line manufacturing met these objectives by using mechanical machines, mechanical transfer line and cam like control system. By its nature there is no in-process inventory at all. The line is dedicated to producing one specific component, in an optimized way.

The advantage of a dedicated transfer line is that it is optimized to produce a large quantity of a single component; it incorporates only the operations necessary for the production of that particular item. The line must be designed for a predicted peak volume, although the pattern of a product cycle is that it ramps up gradually to the peak period followed by a gradual decline. Even if the predicted peak production volumes are realized, a dedicated line remains underused during ramp-up and ramp-down.

The disadvantage is that it requires a major capital investment up-front, and quit a long time to design, construct and test the line. Changes are inevitable, and making changes to the process or components will stop the line production and will require substantial amounts of cost and time.

To reduce processing cost it is desirable to have the advantages of automated line of minimum WIP, but without the dedication to one component, to keep it flexible, to be able to accommodate a variety of components.

2. FLEXIBLE MANUFACTURING SYSTEMS - FMS

Being in the computer field, said Mr. CC, there are a lot of publications on computerized manufacturing systems - CMS. The main theme is to have a system equivalent to the mechanical line manufacturing but by computerized machines and computer control. Many processing machines are computer controlled, the programming of the operations is done by computers, so why not connect control the transfer line machines by computer. It makes sense, and thus the same advantages achieved by automatic mechanical line manufacturing may be achieved by computerize manufacturing.

The essential physical components of CMS are:

1. Potentially independent numerically controlled (NC) machines.
2. A conveyance network to move parts and sometimes tools between machines and fixture stations.
3. An overall control network that coordinates the machines, the parts-moving elements and the work pieces.

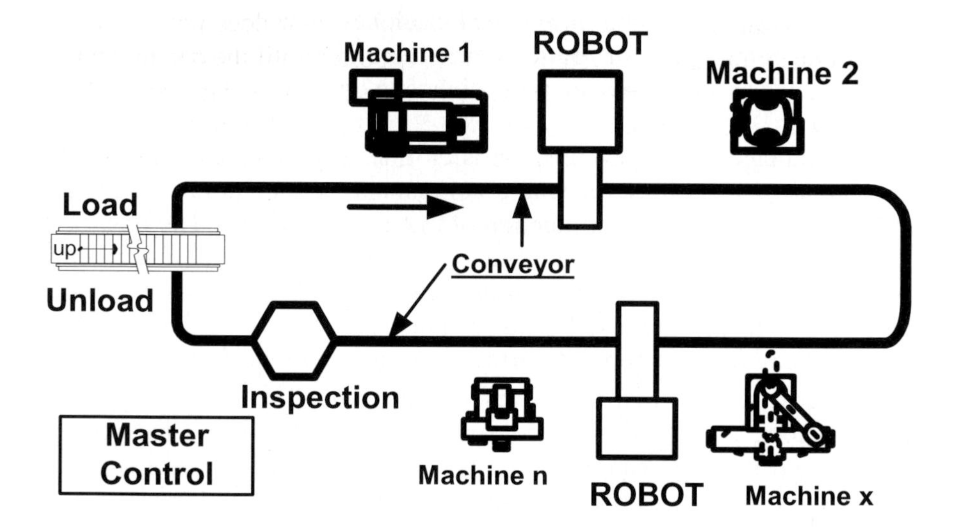

Figure 4-1. Schematics *of Flexible Manufacturing System*

Fig. 4.1 shows schematically FMS system. Item flow begins at the load/unload stations, where the raw material and fixtures are kept. The Master Control computer keeps track of the status of every item and machine in the system. It continually tries to achieve the production targets for each item type and in doing so tries to keep all the machines busy. In selecting items to be sent into the system, it chooses item type which are the most behind in their production goals, and for which there are currently empty fixture/pallets or load stations. If an appropriate pallet/fixture combination and a work piece are available at the load station, the loader will get a message at his computer terminal to load that item on its pallet. He will then enter the item number and pallet code into the terminal, and the computer will send a conveyor (transporter) to move the pallet to the appropriate machine.

Once in front of the machine, the computer actuates the transfer mechanism (robot) in the queue and the pallet is shifted from the transporter onto the machine. The item and pallet wait until the item currently being machined is completed, and then the two items and their pallets exchange position. As the new item is moved onto the machine, the proper NC part program is downloaded to the machine controller from the master control computer. After completing the downloading, machining begins.

The finished item waits for the computer to send a free transporter to collect it and carry it to its next destination. If, for some reason, the part cannot go to that destination next, the computer checks its files for an alternate destination. If one exists, the computer decides if conditions in the FMS warrant sending the item to that destination. If it does not, the item either circulates around the system on the transporter until the destination is available, or the transporter unloads it at some intermediate or storage queue, or retrieves it when the destination is available. The last destination is usually the load station, now functioning as an unload station where item is removed from the pallet.

Mr. C noted that the CMS name was changed and it is called FMS – Flexible Manufacturing System. The emphasis is on the results and not on the technique. The drawback of mechanical line manufacturing was that the setup is for processing one specific item. It was done manually, therefore it took quit some time to adjust the line to move from producing one item to another. With the flexible manufacturing system the objective is to produce medium to low quantities with the efficiency of mass production. It can be done as the setup and controlling the sequence of operations of the machine by computer and it can be done in seconds.

The flexible manufacturing system can be defined as a computer-controlled configuration of semi-independent work stations, and a material handling system designed to efficiently manufacture more than one kind of part at low to medium volumes.

Mr. Q said that he was worried about the quality of the produced items. The computerized machines of the system are very accurate, but the accuracy of the product is a function of the tool setup, the holding jigs and fixtures in the machines and the clamping accuracy. In FMS it is done automatically, and probably the position of the item on machine bed might not meet the accuracy required.

Mr. PM responded that Mr. Q was right. It is almost impossible to clamp automatically a variety of items on the machines and keep the accuracy. Therefore the solution is to use pallets. The idea is that the machine clamps a pallet and not an item, i.e. from the system view point there is only one defined item to clamp. In such case there is no problem to keep the accuracy. The problem is now to clamp the item on the pallets, and this task is done manually at a separate station in the tool room.

The main idea of the computerized systems is to move as many tasks from the shop floor to the office, and thereby allow the production to flow with the minimum of interruptions.

Mr. PS noted that theoretically FMS is a solution to the desire to have a flexible system that allows the processing of single or low quantity items with the efficiency of mass production. However, in practice it does not meet expectations. The main problem is that the number of items on the conveyor (the transporter) is finite. It is constrained by the length of the conveyor. Thus the master control can load only items that are on the conveyor. Usually working with FMS does not call for line balancing. It might occur that items waiting for the next processing operation will have to circulate around the system on the transporter until the required machine is available. Thus the system might become blocked; no new items may be served.

Mr. D interrupted and noted that FMS system may take many shapes, it does not have to be like the scheme as shown in Fig. 4.1. The objective of eliminating work-in-process is worthwhile but wishing for the "best" restricts a "good" solution. It is possible to add a small amount of buffers in the system; it will enable a small amount of WIP but will improve significantly production efficiency. FMS may be constructed as a line manufacturing with buffers in parallel with the machines. It may add bypasses to the circular system to allow items to use shortcuts and thus increase transportation time. It is possible to set the machine layout in a circle with a robot in the middle to serve as a transporter. It is possible to use an AGV as a transporter, which allows using machine layout without any constraints.

The president interrupted and asked for clarifications as to AGV.

Mr. PM explained: An Automatic Guided Vehicle (AGV) is a driver-less industrial truck, usually powered by electric motors and batteries. Modern AGVs are computer-controlled vehicles with onboard microprocessors. Most AGV-systems also have system management computers, optimizing the AGV utilization, giving transport orders, tracking the material in transfer and directing the AGV traffic.

Automatic load handling is used in many AGV-systems. The AGV can pick up and drop off pallets or transfer loads automatically using fork attachments, conveyors, lift tops etc. AGVs can be equipped with robot arms and grippers and perform robotic handling functions.

Several methods of guidance and navigation can be implemented. The early AGVs tracked an inductive guide wire or an optical visible line, painted or made with tape on the floor. The inductive guide wire is still the most used guiding system for AGVs running on concrete floors. In later years AGV guiding and navigation systems with laser scanners, microwave transponders, inertia gyros, ultrasonic sensors, embedded magnets, camera vision systems etc. have been launched. Some modern guide wire systems have the guide wires only as static guidance reference and the AGVs can do many moves off the wire. Modern AGVs commonly use radio communications with FM-radio to transmit data to and from the vehicles. The radio communication by itself has created a lot of freedom in modern AGV system design.

The question of AGV navigation is always a question of the degree of freedom, and shall be taken in consideration in relation to system price and the costs for changes and maintenance of an AGV-system down the road.

The drawback of many free-range navigation systems is that the software gets extensive and may become hard to maintain and change for other people than experts.

Using AGV may allow storing items on shelves and when needed to send the AGV to transport them to the appropriate machine. This of course increases WIP but also increases flexibility and efficiency.

Mr. C noted that FMS was intended to produce items on-line with their assembly, i.e. one of a kind instead of in batches. This calls for a production line starting with raw material and ends with a finished product. Breaking the line between item processing and assembly calls for having a buffer between these two departments and increases WIP almost as in batch manufacturing.

Mr. PS referred to the problem of production planning and scheduling in an FMS environment. Traditional production planning allocates items and operation to processing resources, while ignoring the question of how the items will arrive to the resource. It assumes that any item within any quantity will be available at the right location. It may schedule several items without being concerned that they will not be available for processing. Production planning with FMS it is a different story. Scheduling may load only items that are on the transport mechanism, which is a limited number of items. Therefore the production planning must be divided into two separate stages, stage one plans the items that will be loaded on the transporter, while stage two schedules what items to load on the resources. Experience has shown that the sequence of items loaded at the loading station, has an effect on the efficiency of the system. With a different loading algorithm the throughput may vary more than 100%. Thus the FMS may be very efficient, but also very inefficient.

Mr. F noted that moving toward the FMS system is an expensive endeavor.

The president summed up the discussion and noted that FMS sounded like a very advanced manufacturing method. It calls for in-depth planning and design in order to fully utilize the system, and requires a large investment. Yet in our position we cannot allow ourselves this luxury.

Mr. C proposed to evaluate using FMC system.

3. FLEXIBLE MANUFACTURING CELL - FMC

Mr. C explained that FMC is a set of machining center and conveyor.

Mr. F asked for clarifications of what a machining center is.

Mr. PM explained: a machining center is a CNC – Computer Numerical Control machine, which is capable of performing all operations to a ‘family’ of parts, components or complete products. The machine center is like a mini-factory within the factory which can manage its own operation. The machine can load and unload items from its bed to pallets exchangers. The machine can store a large number of tools and has a built-in tool exchange mechanism. Thereby the long and costly changeover times are minimized. Machining heads have control over 5 or 6 axes of motion, therefore, it has access to all item surfaces, excluding the ones that are chucked on.

Mr. C took over and continued explaining what the FMC system is. Fig. 4.2 shows a schematic representation of the system.

The main features are the machining center machine and a conveyor. Items scheduled for processing are loaded on pallets. At the load/unload station they are mounted on the conveyor, while at the control station their name, position, and CNC program are recorded.

The conveyor moves parts to the pallet exchanger on the machining center and waits. When item processing is done, the machining center turns the pallet exchanger by 180 degrees, thus moving the finished item to the conveyor, and the material for processing to the machine bed. At the end of this exchange, the CNC program is loaded into machine control to start fabricating, and signals the conveyor to move till the next scheduled item is at the machine load/unload station.

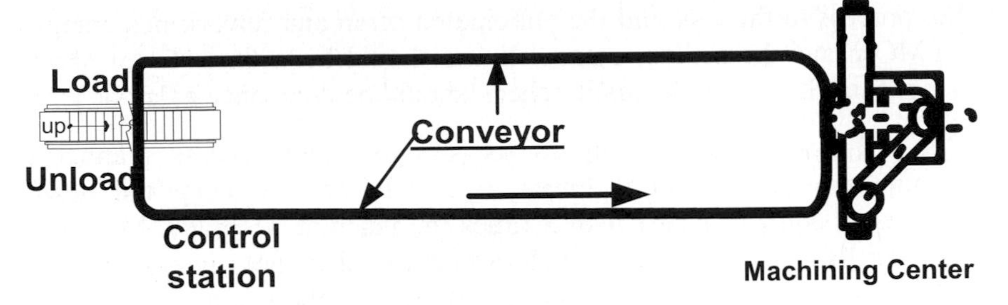

Figure 4-2. Schematics of flexible manufacturing cell

The benefit of this system is that an operator may load the conveyor with all scheduled items, press the push button, and move over to perform other tasks. The FMC works unattended. It is possible to operate a second and/or third shift unattended.

The president was impressed by this system, however, he was concerned about the quality, if the system works unattended who takes care of the quality of items and products? Can an inspection station be added?

Mr. PM responded that an inspection station can be added, but at cost. To make sure that the quality is secure, only items that pass process capability tests will be allowed to be produced by FMC system.

Mr. CC said that he is not familiar with the term "process capability", can anyone explain what this is?

Mr. Q explained; "Process capability" is a term used by SPC - Statistical Process Control. SPC is a technique for error prevention rather than error detection. SPC products will be of the required quality because they are manufactured properly and not because they are inspected. Thus it increases productivity by reducing scrap and rework and provides continuous process improvement.

SPC is accomplished by technological means, statistics for detection and technology for prevention.

Process capability is the measure of a process performance. Capability refers to how capable a process is of producing items that are well within engineering specifications.

A Capability study is done to find out if the process is capable of making the required items, how good is it, or if improvements are needed. It should be done on selected critical dimensions.

A paper on SPC is attached in the appendix.

Mr. PM continued to answer the question of quality. The process must be in control in order to be qualified as "process capability". A process in control has its upper and lower control limits which establish the suitability of the process to the task and the anticipated scrap and rework percentages. For FMC usage the process control limits should be within ±6σ or above, which means that a sample quality check should be done once a day or 1:25.

Mr. PP noted that it is up to the designer and process planner to determine the desire control limits. It is advisable to keep the design tolerances to equal or larger than 3 times the machine accuracy. Moreover, the slogan that quality increases processing cost does not always apply. In many cases, with good and accurate machines like the machining center, it will be difficult to produce a rejected item.

Mr. PM disagreed with the last statement of the process planner.

Mr. PP stood by his statement and explains; the design tolerance is what the designer allows for the functionality of the product and ease of assembly. However, the process plan and the machine accuracy will determine the accuracy of the produced item. It has nothing to do with what is specified in the drawing. The process planner is bound by the design specifications and tolerances, but in many cases the need to meet one feature affects the result of another. For example; the drawing specifies a blind hole with rough tolerances. However to process the blind hole a reamer or end mill must be used. Hole diameter will be controlled by the tool and not by the design specifications, which in this case is tighter than the specification.

Mr. F is impressed by the capabilities of FMC, particularly by the ability to work unattended. This feature enables to run the factory in three shifts while employing personnel for one shift. Working three shifts will increase production, will reduce lead time and thus reduce WIP. The second and third shift do not need light, air conditioners, meals, etc. thus additional saving will be achieved. Although I like FMC system I recommend that a return on investment - ROI study will be made to determine if it is worthwhile for our factory.

Mr. PP noted that although FMC sounded promising, it was necessary to check if it was really possible to work two shifts unattended. The conveyor must accommodate items for these working shifts. If we assume that a pallet size is 400 x 400 and an average machining time of item is 10 minutes, then to work 16 hours unattended the number of pallets to be loaded on the conveyor is: 16*60/10 = 96 units. To store that amount of pallets the conveyor length must be: 96*0.450 = 43 meters, hence it will occupy space of about 40 square meters not including the machining center space. Such space should be taken into account in evaluating ROI.

In a sense FMC might cause waste of power and processing time. What I mean is that machining centers have motor power of about 35 KW. While processing time of a "normal" item needs about 5KW.

Rough operations require power, while the finishing operations require speed. Average power and machining time to machine one surface of 300 x 100 are:

Rough milling	20KW	0.38 Min.
Finishing milling	2.7KW	0.99 Min.
Drill Ø10; 20 mm long	0.5KW	0.24 Min.

This means that the power of a machining center is utilized by no more than 33% of the time. It will be much wiser to have the finishing and hole making operations on a separate machine.

Mr. PM rose angrily and noted that such arguments are out of place as the waste in machining is compensated by gain in production planning and scheduling. Machining an item in one or at most on two stages increases accuracy and prevents mix ups in production control.

The president summed up the discussion: FMC is an impressive system. However, taking into account the points made and our present financial position it calls for careful analysis to see if it is appropriate for us.

Mr. PP noted that there is a manual system that gave the inspiration to the FMS and FMC systems. It is Group Technology, and its modern version is called Cellular Manufacturing.

4. CELLULAR MANUFACTURING AND GROUP TECHNOLOGY

The President asked Mr. PP to introduce these systems.

4.1 Group technology

Mr. PP explained: Group Technology (GT) is a manufacturing philosophy aimed at increasing productivity in manufacturing of the job-shop type. The main goal of GT is to produce a single or small quantity items with mass production efficiency.

Mass production systems eliminate work-in-process, minimizes manufacturing lead time (throughput) and quickly detects quality problems. To achieve it a transfer line dedicated to produce one specific component, in an optimized way is constructed. The raw material enters the first working station, and a finished product exits at the last station of the line.

GT proposes to achieve similar results by constructing work cells.

Mr. C asked for clarification as to what a "work cell" is.

Mr. PP explained: A work cell is a machine layout in which raw material enters the cell, and a finished product exits the cell. Hence, one work cell might include diverse sorts of machines, fixtures and tooling required to produce a family of parts. It is different than the usual functional layout of equipment with no interrelation between groups of different functions. Each part takes an unpredictable path through the shop in order to reach all the necessary machines involved in its processing. Every time a job is moved from one (operation) workstation to the next, there is a delay. Production control becomes extremely complicated and it is almost impossible to get realistic up-to-date information on the production status of any particular job.

Mr. C asked for clarification as to what a “family of parts” is.

Mr. PP explained: A family of parts is defined as a collection of related parts that are nearly identical or similar. They are related by geometric shapes and/or size and require similar machining operations. Alternatively, they may be dissimilar in shape, but related by having all or some common machining operations. Parts are said to be similar in respect to production techniques when the type, sequence and number of operations is similar. This similarity is therefore related to basic shape of the parts or to a number of the shape elements that are contained within the part shape. The type of operation is determined by the methods of machining, the method of holding the part, tooling required.

GT work cell is manually operated. Usually the cell is composed of about 5 to 7 machines. As the machines are layed out close together one operator may attend several machines. The transfer of items between operations is done one piece at time (not batch) by hand or by buffer between machines.

GT is concerned with the lead time and thus the work-in-process. It is claimed that in batch type production only 5% of the lead time is direct working time, whereas 95% of the lead time the part waits in the shop to be processed. Furthermore, the 5% can be divided into 30% actual machining time and 70% for positioning, chucking, gauging, and so on. Hence, only 1.5% of the lead-time is actual machining time. GT directs its effort toward reducing lead-time by attacking the remaining 98.5%. One way to achieve this is by organizing the plant layout according to work cell rather than functions.

With GT work cells the saving will be in transfer time between operations and reduced set up times and reduction of work-in-process, as the transfer of items between operations is one by one as if it works with automatic transfer line. The batch size for a family of parts will be the sum of all parts of the family, thus increasing the number of parts per setup and thus reducing set up time considerably.

The cell is regarded as a machining center, where raw material is the input and the output is the machined product. Thus it drastically reduces the scope of production scheduling and control problems.

The cell workers are responsible to meet scheduling and for the quality of the products. Production control becomes extremely simple and it gets realistic up-to-date information on the production status of any particular job.

The president was very impressed with GT. As explained it will reduce cost and lead time and WIP thus it is exactly what is needed, and it seems that it will require modest investment, so let's go for it.

Mr. CC, supporting the president's enthusiasm, said the philosophy of GT is excellent. Actually the goals and applications of GT are expanded beyond that of work cell manufacturing technique, and the broad meaning of Group Technology now covers all areas of the manufacturing process. Some of them are: Process Planning, design, Material management and purchasing, Material management and purchasing, Cost estimating.

Savings in process planning result from using the same process for a family of parts. Examining the actual process plans in a shop usually reveals that for similar parts, belonging to the same family, many different processes are on company files. This can be explained by the fact that several process planners were involved in this task, or that they were made at different times. GT proposes to examine the different process plans and evaluate them in order to find the "best" process. This process will be the master process plan. It is suitable for a virtual part of the family. The specific part will retrieve the master process plan and update it to suit the specific of the specific part. By applying the master process plan to the available part, immediate improvement and benefits will be achieved. When a new processing technique becomes available, the master process plan will be updated.

In creating a new part design there is the design time, detail drafting time, prototyping, testing, and documentation and certainly drawing maintenance. When the new part design hits manufacturing many things happen. There is tool design. Tools have to be either made or bought. Time study is involved. Production control has to schedule the part, cost accounting is involved, data processing, purchasing, quality control, N/C programming are all affected – we could go on and on. It is expensive to support new parts. With GT technique some of these expenditures can be avoided.

The GT concept is to carefully examine the active parts of the company, and create families of products and parts and make them company standards. When a new part is required, before rushing to design, GT retrieves and compares the available parts to decide if they can be used. It has been reported that from 5 to 10% of the annual output of new parts could be avoided by the proper use of family of parts technique. A company can save from $250,000 to $500,000 just by reducing the duplicate design. This is without making anything.

Saving in cost estimating will be achieved using past estimates. Determine to which family of parts the new parts belong. Retrieve the cost of the master part cost and perhaps add a factor and arrive at estimated cost. Experience shows that a very accurate cost is determined.

The use of a group of materials has lead to greater purchasing efficiency, lower stock levels, and savings in procurement. GT using a family of parts may reduce the number of orders through blanket orders and through larger lot sizes. Parts are bought on a 'family of parts basis'. Blanks may be purchased to suit a family of parts and not any specific part. It might increase processing time, but reduces purchasing and inventory expenses, and probably lower blank cost.

Mr. F went on summarizing GT philosophy by a few words: "Do not invent the wheel over and over again". But declaring and doing is two separate things. Since 1950 Group Technology was promoted heavily in several countries but unfortunately only few factories succeeded to implement it.

Mr. D: It is true that with all the energy and research GT did not prove to be an all round success story. However, there were a few factories that did exercise GT with good results and reports on its use inform us that it is "immensely successful,… there was an average savings of 70% of machine time, Work in progress reduced by a ration of 8 to 1". There is no reason to doubt the accuracy of these reports or that that GT is an excellent technique for certain type of industries. The question is why the GT system was not widely practiced and hence was abandoned (or being used without knowing that they practiced GT). The answer is that in order to practice group technology as a systematic scientific technology, tools for the identification and creation of part families must be prepared.

There are three basic methods to form a family of parts, namely:

- Manual – walk around the shop and look.
- Production flow analysis
- Classification and coding system

Many of the reports on successful group technology applications have come mostly from studies in which the main work on the manufacturing concept was done with families of parts that had been organized manually. Engineers have tended to view each part produced in the company and make a human decision, relying on their memory and on the flexibility of the human mind. Therefore, this method is excellent for small companies were the human mind might grasp all the part produced in the company.

Mr. CC noted that as GT's wider application is in design, process planning, work cell etc. it seems that each one of the applications requires a different classification and family of parts.

Mr. D asked why do you think so, why cannot one family serve all applications?

Mr. CC explained: for example in your area, design, when a new part is required, it is recommended to retrieve and compare the available parts of the family to decide which one is the most suited to be used or modified to the one required. Remember, a family of parts, is a family with individual parts. Suppose that in retrieving "the most suited" you get 2500 candidates? It will be of no use, as the time to examine such a large number of drawings might exceed the time to design a new one. For this application the size of the family must be small. However for process planning or for work cell design, the size of a family of parts must be as large as possible. To form one system that will be design oriented, production oriented or resource oriented is practical asking for the impossible.

Mr. PM said that he might understand the term 'family of parts'. However, GT calls for family of fixtures and tooling. Although the parts in a family have some similarity, yet they might be of different dimensions. A fixture must hold each member of the family in the right position; otherwise the accuracy could not be met.

Mr. D explained that it is possible to design tooling and fixtures by using group concepts common to the part family. To use tooling and fixtures to the full, the operations must be arranged so that the maximum number of parts in the family can be processed in one set up, which means that jigs accepting all the members of the family have to be designed. For example, the design of a master jig with additional adapters is one way of dealing with changes in size, number of location of features. As the results of these advantages of group technology the cost reduction in tool design, tooling and equipment, production control etc. becomes very significant. By the way, these problems of fixtures are common to FMS and FMC systems as well.

Mr. PM: I can understand your design capability, but still, it seems to me that a work cell is organized to produce a family of parts, what will happen when the product mix is changed. It seems to me that the task of creating families of parts is an endless job; I do not see a big problem with it, as it is a clerical task, but the work cell has to meet the family, this means that plant layout must be dynamic. I do not see how we can move machines around any now and then. Furthermore I see some more problems such as:

Work-cell supervision: In the functional layout, it is required that the foreman and operators be an expert in one type of manufacturing. In the work cell concept, there is variety of resources, in order to operate and to instruct the operators, the foreman must be an expert in several fields.

Work distribution: With the functional layout there was no problem in transferring an operator from one machine to another, since they were of the same family and the operator had the training and skill to operate them. In work cell environment, however, this cannot be done.

Shop floor Control: Since the work cell is built around the most complex work piece-processing route within the group, it is possible that a condition of a single machine overload in one cell and underload in another will occur for a certain machine. Hence, machine utilization and work balance will be low.

Production Planning: A condition in which one work cell is overloaded and another underloaded can occur.

Mr. C was not impressed with the supervision and operator skill problem. All that is needed is training in teamwork as well as manufacturing techniques.

Mr. F agrees that these problems can be resolved by training. However, that can be a time-consuming issue to resolve. Each station or piece of equipment requires varying degrees of skill to operate. This training must be done before the cell layout is designed, because it's very important that the operators are involved in the cell's layout and planned operation. They are the people who know how the equipment operates and understand how to do their assigned jobs. Operators need to understand what cells are, how they work, how they differ from traditional "batch and queue" operations, and the objectives of the cellular environment.

In addition to equipment and team training, operators need training on how to perform setups, setup reduction, inspections, preventive maintenance, proper equipment cleaning procedures, and other such activities.

I think that with all the promises that GT makes we should drop the idea of attempting to introduce it in our plant.

Mr. CC: Group Technology was introduced in the '50s and was based on manual techniques. Introduction of computers techniques enhanced the philosophy of GT. One of the modern versions of the concept of group technology work cell is Cellular manufacturing. I think that it might be of interest to our group, to learn what the features of the modern version are. A short description follows.

4.2 Cellular manufacturing

Mr. CC gave a short description of the cellular manufacturing system; The cellular approach object is that only the amount of product needed by the customer should be produced. It usually requires single-piece flow or, at the least, small batch sizes. The method to meet this objective is to form family of parts, and rearranging plant processing resources, to form manufacturing cells.

The implementation of cellular manufacturing requires the following steps: analyse the open orders for a certain long period, decide upon a product family of parts, determine the operations required in the cellular environment, design jigs and fixtures that will reduce setup time, balance operations between operators, design the cell layout, move equipment to form the cell. Since most modern processing resources are flexible by nature, and they can perform several jobs, it is easier to practice cellular manufacturing than the group technology. The cell might be a virtual cell that will not require the move of resources any time the product mixes, and the orders are changing.

Implementing manufacturing cells affects the production schedule. In many plants today, production schedules depend upon customer forecasts, equipment and material availability, and overdue customer orders. Large batch sizes are run to reduce the number of required equipment changeovers. In cellular manufacturing the batch size can be exactly the quantity of the customer's order. Due to design of modular fixtures and computerized operated processing resources set up is not a hindrance any more.

Production schedules must adapt to the cell's operation. They need to be more flexible in the amount of product produced, and more precise in the amounts of product output.

Inventories such as work in process, raw material, and finished goods are listed as assets on a company's balance sheet. But high inventories are really liabilities that tie up company resources. An operation must introduce methods of reducing raw material, WIP, finished goods inventories, and setup times for a cell manufacturing system to work.

It is advisable that the cellular approach must be applied to the entire production line. Picking isolated areas in which to implement manufacturing cells results in islands of success, but may not allow a product line to become efficient. The company may still depend upon operations that run in the traditional manufacturing environment. If the cell or group of cells doesn't include all operations in a product line/family, a cellular system will have a minimal impact on the overall production process. The cell contains processing resources of several capabilities. Operators have to be flexible as well as the resources in the cell, therefore they have to be able to operate all the resources in the cell, as well as how to setup each resource.

The cellular manufacturing calls for teamwork. The responsibility for quality and meeting due dates as well as internal scheduling is of the group as a unit.

Mr. F was not impressed with the new version of GT, the only new idea is that a work cell can be "a virtual cell that will not require the move of resources any time the product mixes, and the orders are changing." It is a vague statement but if it really can be done it solves only one of the problems, and creates a new one.

Mr. PP: We had the idea that we may look at the plant as one work center (which it is) and devise a routing that will lead the parts in only one way, i.e. the parts enter the plant at one point and proceed toward the end of the plant without zigzagging on shop floor. As a process plan it is possible, but I do not see why it should be done under the name of GT.

The president summed up the discussion: GT is an impressive system. It might meet our present financial position. However, taking into account the points made, it calls for careful analysis to see whether it is appropriate for us.

Chapter 5

WORK-IN-PROCESS IN BATCH TYPE MANUFACTURING

Lead time reduction

1. PRODUCTION MANAGEMENT

Mr. PM opened the session by saying that actually he does not understand the relevance of work-in-process in batch type processing.

Production planning and scheduling controls the activities of shop floor, with the objectives of:

- Meeting delivery dates.
- Minimum idle times on the available resources.
- Reducing manufacturing lead time.
- Keeping to a minimum the capital tied down in production.

The lead time may be regarded as the work in process. Usually when the routing of an item calls for several operations each on different work stations, then item cost value increases gradually. One may refer to it as WIP, but in production planning work-in-process is kept to a minimum by the objective of reducing lead time. As far as I understand by talking to accountants, this WIP is considered as overhead, and not accounted directly to a product.

Mr. F: You are referring to conventional cost accounting systems, where direct costs such as the cost of specific services are billed directly to the product. However, indirect costs such as the WIP are regarded as overhead. In this system, overhead cost per hour is the same irrespective of the job type. The objective of modern accounting systems is to have an accurate cost reporting system, by which management may reduce the risk that poor case-mix decisions, faulty pricing decisions, and suboptimal capital budgeting decisions will be made because of inaccurate costs.

Mr. PM agreed that it was a good cause but in batch manufacturing, where hundreds of items, of different products and at different stages of processing are on shop floor, I cannot see how it can be done. It will probably cost more to gather such accurate information than the possible benefits from using that information.

Moreover, the objectives of production planning conflict with each other. To minimize the capital tied down in production, the work should start as closely as possible to the delivery date; this will also reduce the manufacturing lead time. However, this approach will increase resources idle time in an environment in which resources are not continuously overloaded.

The objective of minimum idle times on the available resources calls for keeping a queue before each resource. To process a part usually requires several operations each on a different resource. Scheduling plans the sequence of operation with an attempt to keep continuity and to keep lead time to be as short as possible. However, the situation on the shop floor is dynamic; disruptions occur, machines break down, rejects are produced and thus keeping the schedule up to the minute is impossible. To avoid machine idleness, each machine must have a queue in front of it, in order to have jobs in case that the next job is late, or that the present job was ended earlier then planned. It is a common practice in scheduling to allow 1 to 2 days between operations, to allow for waiting for transportation, transportation time, waiting in next operation queue. This practice increases WIP, but it cannot be billed to a specific product, but it must be done in order to decrease production cost.

Another factor that produces WIP is the basic working order of the batch type manufacturing. The processing is by batches. The batch is transferred from one workstation (operation) to the next and not to each individual part. Thus each part is waiting till all the parts in the batch are processed. This waiting time is also WIP. That is the nature of batch type manufacturing.

1.1 Controlling batch size

Mr. PS agreed with the above description and that there is nothing to do in order to eliminate WIP. However, it is possible to reduce WIP. The batch size is a factor that determines the WIP, as parts waits till all members of the batch are processed. Reducing the batch size will reduce WIP.

Mr. F interrupted and noted that the batch size is determined by economic considerations. Therefore, such reduction is already taken care of by using the theory of economic lot size - EOQ.

Mr. I agreed with Mr. PM that there is no way for WIP reduction methods in batch type manufacturing. We are very efficient and keeping EOQ to optimum level. We use order point and safety stock techniques to assure minimum WIP while keeping service level high.

Mr. PM interrupted and said that he disagreed with Mr. I. Setting EOQ (lot size) for products is production planning and scheduling responsibility and not that of inventory. It is done in the master production scheduling phase. Manufacturing activities starts with customer orders, or with sale forecast. Customer orders specify: the product, the quantity and delivery date. The order quantity is actually replaced the EOQ.

The master production schedule is a coordinating function between manufacturing, marketing, finance, and management. It is the basis for further detailed production planning. Its main objective is to plan a realistic production program that ensures even utilization of plant resources - people and machines. This will be the driving input for detailed planning and will ensure, as much as possible, against overload and underload of resources at all periods of time. In case that it foresees problems it may ask marketing to negotiate with the customer a new delivery date, and a delivery schedule that will meet production economic lot sizes. But the final decision is with the customer.

Mr. I agrees that the product EOQ is being set by marketing and customers and master production scheduling. However, it is the lot size of the product and not of the dependent items. For those items the method of setting lot sizes by inventory control may be applied.

Mr. PM disagrees, you must remember that lot size in production has nothing to do with purchasing, using order point, or safety stock. Production is in house, and being controlled by production planning system that sets its own lot size and schedule. Scheduling dependent items is done to serve the assembly of the product. If any item is missing for assembly then all other items wait on shop floor (WIP) and the assembly station might be idle, thus increase WIP. The importance of the master production schedule is that it is the key to the success - or failure - of the detailed production planning.

Mr. FM interrupted and noted that master production scheduling is a good starting point for planning activities. However, no one actually believes that the master production schedule will be accomplished as predicted. Since new orders and changes in existing orders occur continuously, machine breakdowns, rejected items produced, etc. have impact on the actual performance. The life of a production schedule depends on the environment in the shop. It is probable that after a short period the schedule is no longer realistic, and thus a new plan is required. When a large number of operations are involved, the problem can rarely be solved satisfactorily. In addition, a satisfactory solution in one area may cause an unexpected problem in another area. Most WIP on the shop floor is not planned but is due to the dynamics on shop floor.

Mr. PM agreed that most WIP on shop floor is not planned, yet it must be planned in order to aim for minimum WIP. Lot size for the requirement planning method is calculated differently than the EOQ.

Requirement planning is designed to establish order due dates that correspond with the exact date of need, that is, when assembly is scheduled to begin. No safety stock is required, since it has been accounted for in the forecast and master production schedule.

The calculation of the lead time for manufactured items is based on routing data and modifiers, such as transport time, queue wait time, and inspection time. The formula is unique for each item, and no extra safety lead time is required. The timing of replenishment orders, as calculated by requirement planning, is based on the master production schedule, on the one hand, and on the lead time, on the other hand.

Mr. FM said that this algorithm is very impressive; however, it is not realistic, as the number of periods is based on planning with infinite capacity. The requirement planning schedule, with its infinite capacity, serves mainly to set objectives for capacity planning and purchasing; Moreover, it schedules backward from the product due date and hence requirements may fall into the past for example, and requirements are overdue. In requirement planning, there is no automatic feature to shift the product network by forward scheduling, starting from the current date.

Mr. PM agreed with these remarks. The planned orders of requirement planning represent no commitment, since no action has yet been taken on them; they are an input to capacity planning and job release for the initiation of action. Hence, the lead time is for information purposes and there is no need for such safety factors as safety lead time. But that is the best that can be done.

2. WORK IN PROCESS DURING PROCESSING

Mr. C noted that our method of capacity planning and scheduling is not flexible and cannot comply with the dynamics of shop floor, this is one source of accumulating work in process. Another source is created by job waiting at the resource, waiting for setup to be completed. I propose to discuss these two areas.

2.1 Scheduling methods effect

Mr. FM: Scheduling faces the problem of jobs competing over capacity. When a resource becomes idle a decision must be made about its next job, considering only those jobs that are in the queue of that resource.

There are many rules (dispatching rules) to guide the scheduler as to which job to load first. Some such rules are:

FCFS -	First come first served
DDATE (EDD) -	Has the earliest due date
RANDOM -	Select randomly
SLACK -	Has the smallest ratio of slack time to number of remaining operations
SPT -	Has the shortest processing time
SIMSET -	Has the shortest actual set-up time

The question arises as to whether any one rule works better than some other and whether any decision rule is significantly superior to another. This leads to another question: How do we measure and define good scheduling? What are we actually trying to accomplish? There are many criteria by which one can define the goals of scheduling, such as:

- Minimum level of work-in-process,
- Maximum number of jobs sent out of the shop,
- Maximum number of processes completed,
- Minimum number of processes completed late,
- Minimum number of jobs waiting in shop,
- Minimum queue wait time of jobs in shop,
- Maximum shop capacity utilization.

They are all important objectives, but you cannot have them all.

The most favorite rule is the SPT; it makes sense as the shorter the processing time, the more jobs can be done.

I propose that our production planning will use the SIMSET rule. It reduces setup time and thus increases the utilization resource time, reducing work in process.

Mr. PS agreed that the SIMSET rule would reduce setup time. However, it is not practical as production planning file do not hold data on setup. Thus with all good will it cannot be done.

Mr. PP does not agree that setup data is not available. Process planning, as I do it, includes setup definitions and routing details. It seems that some data is lost in the system.

Mr. PS: I am sure that someone defines the setup details that go with the routing. What I get in my files is job processing time, resource number, and the time to setup the resource for that operation. There are no details as to what the operation details are, and no details on the setup. We are using the most advanced computer programs such as MRP and ERP. We do not develop such programs but get what is on the market. Actually we do not know exactly what are the algorithms used in scheduling programs.

Mr. CC supports Mr. PS in his statements. In search for a scheduling computer program, we are doing benchmarking comparing different programs, and with consulting with production planning and scheduling we select the most suitable program. We are in no position to develop our own programs; we must use what is available on the market.

Mr. FM Production planning is doing a good job in scheduling and load balancing. That schedule is transferred to shop floor as a list of jobs to be performed, their sequence, resource to use, setup and processing time of each job. The sequence, as far as I know, does not consider similar setup priority. In practice, due to shop floor dynamics, changes have to be made. In such cases I allow myself to deviate from the planned sequence and as far as I can I give priority to jobs that can use the resource with tearing down and setting up the setup, or adjusting the setup to fit a job that is waiting in the queue.

Mr. C: I would like to propose that in case that it is not possible to use the SIMSET scheduling rule in the advanced computer programs, not to schedule sequence of operation at all. Let production planning, considering delivery dates, prepare a list of jobs that should be done at a period, and leave the detail scheduling to the foreman.

The president concluded and said that he supported Mr. C's proposal, even though he was not enthusiastic about it: I prefer to have a production planning computer program that considers this scheduling rule. I ask Mr. CC and Mr. PM to check this point, and find out the reason for such avoidance. Furthermore, please check if SIMSET rule is really such a good one.

Mr. PM: Listening to the discussion concerning shop floor scheduling, and the president's summary, allow me to outline a brief description of the some systems that follow this line of thoughts. It seems that the idea to let the foreman and fellow workers run the shop floor by objectives and not by accurate scheduling of activities has been adopted by several researchers.

2.1.1 Common-sense manufacturing

Common-sense manufacturing's (CSM) objective is to regulate work in process, and enables the manufacturing line to meet the production goal. It allows operations teams on shop floor to regulate and adjust the working plan.

MRP systems approach the production control task from the "first plan the work and then work the plan" viewpoint. Unfortunately, such systems are often better at planning than they are at working. At the point of actual production, the execution methodologies of JIT systems, such as pull systems and kanbans, are better utilized.

The CSM system uses trays or work holders (called totes) to gauge lot size and to control the work in process. A tote system is a method of handling parts and assemblies during production. It is also a method of tracking lots through the line.

Each area of the production line is analyzed to determine the correct tote and the proper lot size. Many factors may influence a decision on lot size. The ideal is usually a lot size of one part. While this would be advantageous for inventory and interval reduction reasons as well as for lot traceability and tracking, it is often not feasible for other practical reasons.

The first factor in selecting lot size is often the number of parts that are easily processed together as a batch. Other factors include the production facility size and capacity, the physical size of the parts, and the time required to work on a tote full of parts. Often the lot size is set by the constraint operation after taking into account the run time, setup time, and machine utilization factors.

Planned amount of work-in-process inventory is kept as a strategic buffering. This inventory is there to allow for production problems such as breakdown maintenance. It is there, also, to ensure that the constraint operations always have work available, thus keeping them running. The extra inventory also allows improved responsiveness by the product line to short-internal orders or other unexpected demands. It also affords the opportunity to occasionally perform experiments on the line with the production facilities for such things as process improvement. This enables continuous improvements in yields, interval reduction, and costs of manufacture.

Mr. PP noted that the term "production line" was used in the description. We are not line manufacturing, rather we are batch type manufacturing. I am not sure that our products are fit for line manufacturing, and besides, constructing a line calls for a different layout than what we have now.

Mr. PM responded and said that in CMS "production line" does not mean "line manufacturing". One of the things that make the CSM system beneficial is that it does not dictate the organizational structure of the manufacturing plant. The structure that is in place does not need to change as a result of the implementation of the CSM process.

Mr. C noted that CMS is in the line of our proposal to let the foreman have more freedom in scheduling jobs in his department. However, beside CSM there are some other proposed systems that I think that we should evaluate.

2.1.2 Kanban system

Kanban ("tag") is a production planning and scheduling system based on pull instead of push system. The goal of eliminating waste is also highlighted by kanban. Kanban is a powerful force to reduce manpower and inventory, eliminate defective products, and prevent the recurrence of breakdowns.

A kanban is a tool for managing and assuring just-in-time. Kanban is a simple and direct form of communication, always located at the point where it is needed. In most cases, a kanban is a small piece of paper inserted in a rectangular vinyl envelop. On this piece of paper is written how many of what part to pick up or which parts to assemble.

Kanban is a Japanese word that means "visual record" and refers to a manufacturing control system developed and used in Japan. The kanban, or card, as it is generally referred to, is a mechanism by which a workstation signals the need for more parts from the preceding station. The type of signal used for a kanban is not important. Cards, colored balls, lights and electronic systems have all been used as kanban signals. A unique feature that separates a true kanban system from other card systems, such as a "travel card" used by most companies, is the incorporation of a "pull" production system. Pull production refers to a demand system whereby products are produced only on the demand of the using function.

Kanban always moves with the needed goods and so becomes a work order for each process. In this way, a kanban can prevent overproduction, and prevent largest losses in production.

Kanban, in essence, becomes the automatic nerve of the production line. Based on this, production workers start work by themselves, and make their own decisions concerning overtime. The kanban system also makes it clear what managers and supervisors must do. This unquestionably promotes improvement in both work and equipment.

The main characteristic of kanban is its operation simplicity, and its ability to reduce work-in-process. It is based on working to buffers. Buffers exist to protect the system from delays in production. Buffer size, however, is a trade-off between protection and lead-time. If the buffer size is increased, the protection increases, but so does the manufacturing lead-time.

Once a kanban-activated workstation has filled its output buffer it is not authorized to produce output again until the output buffer is depleted to its reorder point. The workstation is said to be "blocked."

Mr. FM: Kanban requires a buffer of material for each possible part in front of each resource. Therefore, for multi-product environments kanban requires substantial inventory to achieve the same throughput. It takes time and effort to compute the buffer size; therefore I believe that it is a good system for repeatable parts, and not for random order parts.

Mr. PS noted that Kanban is a tool for just-in-time realization. For this tool to work fairly well, the production process must be managed to flow as much as possible. Other important conditions are leveling production as much as possible and always working in accordance with standard work methods.

Mr. PM: Some Kanban rules are as follows:

1. Earlier process produces items in the quantity and sequence indicated by the kanban.
2. Later process picks up the number of items indicated by the kanban at the earlier process.
3. No items are made or transport without a kanban.
4. Always attach a kanban to the goods.
5. Defective products are not sent to the subsequent process. The result is 100% defect-free goods. This method identifies the process making the defectives.
6. Reducing the number of kanban increase their sensitivity. This reveals existing problems and maintains inventory control.

Mr. CC: The Kanban system is most likely to be associated with just in time (JIT) system and Theory of constraints (TOC).

The success of kanban systems appears to depend heavily on complete implementation. Even in cases where the implementation is complete, kanban systems are unable to cope with variety and demand fluctuation. It may be that when kanban is used as part of a continuous improvement program, like in JIT philosophy, it is likely to produce increased benefits to the user.

2.1.3 Constant work in process - conwip

The objective of constant work in process is to reduce inventory level and controlling production planning and scheduling.

CONWIP is a closed production management system in which a fixed number of containers (or cards) traverse a circuit that includes the entire production line. When a container reaches the end of the line the finished product is removed. The container is then sent back to the beginning of the line where it waits in a queue to receive another batch of items. During each container's cycle all items in the container are of the same type.

The amount of material put into the container is set by a predetermined transfer lot size.

Since CONWIP systems are closed manufacturing systems, as is Kanban, they have the following advantages over open systems: easier control, smaller variances, and smaller average work in process (WIP) levels (and thus also shorter flow times) for the same throughput. They are also self-regulating.

In addition, CONWIP systems have the following advantages over Kanban:

1. They are very robust regarding changes in the production environment and are easier to forecast
2. They easily handle the introduction of new products and changes in the product mix
3. They cope with flow shop operations with large set-up times and permit a large product mix
4. CONWIP systems also yield larger throughput than Kanban Systems for the same number of containers

Work in process ensures continuity of production by buffering the bottleneck resources.

Mr. FM: Often a manufacturing line does not sit in isolation, but rather is part of a larger manufacturing environment. Just as a machine processing time variance can cause a fast machine to become the bottleneck from time to time, a high variance can cause the CONWIP line to become the bottleneck in the overall system.

Mr. PM: Since CONWIP systems can be viewed as closed queuing networks; one may mistakenly view the system as a loop (having neither beginning nor end). This allows one to 'cut' the line at any point in order to evaluate its performance. This approach, as recognized by the model, is valid for mean performance measures but very unsuitable for the variance of the performance measures.

2.1.4 Cycle time management - CTM

Cycle Time Management is a manufacturing philosophy that is dedicated to reducing inventory and waste. Respect for workers is the vehicle that promotes continual improvements. For too long factory workers have been misguided, misused, mismanaged and thought of as drones. Worker involvement in all aspects of CTM leads to manufacturing excellence. Manufacturing excellence is looked upon as a strategic advantage for achieving global competitiveness. Manufacturing excellence produces a product that meets or exceeds the customer expectations at a competitive price delivered to the customer on time. Manufacturing excellence is much more difficult than buying the latest automated technology. Automated equipment, such as machining centers, is not cheap and has proved to be difficult to debug. CTM may offer the best of automated systems and respects the workers.

Mr. I: The main driver of CTM is inventory reduction. In the past, inventory has been thought of as an asset, a security blanket for achieving productivity. CTM contradicts this belief and simply states that inventory is evil. Inventory hides problems such as design problems, machine downtime, long set-ups, absenteeism, defective parts, poor vendor quality, and past due dates.

Mr. PM: Reduction of inventory through the utilization of small lots and pulls operation exposes the problems and gives workers the opportunity to solve control process problems. These improvement opportunities allow shop floor workers, their supervisors, production engineers, and design engineers the opportunity to work together to solve problems and conduct process refinement activities. The potential for breaking down departments walls with these process refinement activities is great.

2.2 Set up reduction time

Mr. C noted that one of the major contributors to work-in process is the setup time. During setup the resource is idle from a job processing standpoint. In a way this time is a waste.

Mr. PM rose angrily and said that some may call it waste but without setting up a machine with fixtures, tools, adjustment that are required for the specific job no machine can perform any job. Setup is part of the processing cycle.

Mr. C: I would be glad if we can find a way to work with no setup, but I will be happy with reducing setup time. Actually in our previous discussion it was proposed to schedule tasks by similar setup. 'Similar' does not have to be 'exact', but the setup time might still be reduced.

Mr. FM: There are two kinds of setup operations; an internal and external. Internal means that the machine is idle while performing the setup. External refers to setup operations that can be done in a tool room and not on shop floor. An example of using external setup can be in CNC machines, were the machine preparation (setup) is done in the office and does not cause idle machine time, in the use of pallets in flexible manufacturing system (FMS). We may check if some of the internal operations can be changed to external ones.

Mr. Q It is a good idea to have external setup. I just worry about the accuracy of the setup and the product. CNC is an accurate and costly machine, and this method proved itself. However, I wonder if with universal machines and old machines this method can be used.

Mr. D: Jigs and fixtures are built as an accessory to part processing, and their task is to hold the part firmly and accurately in the machine in order to assure item quality. In many cases, especially when working with small lots, its cost might be higher than the machining cost itself. To reduce both the direct cost of jigs and fixtures and the cost of setting those up a modular fixturing system can be used.

Mr. F asked what do you mean by modular jigs and fixtures, and how can they reduce cost?

Mr. D explained: A modular fixture is a method of designing a fixture that can be used for a group of parts. There is no need to design and build a special jig for each part. The fixture is composed of building blocks assembled in a suitable manner. For example: a base plate has T-slots or locating holes machined at right angles. The slots are parallel to each other to ensure the accurate alignment of fixturing elements. Large structural elements such as angle plates, clamping plates, and blocks are used and have the same T-slots to align, locate and attach other elements of the set. Precision position holes along with tapped holes to align, locate and secure fixturing elements are also parts of the modular fixturing system. The individual blocks are very accurate: figures of 0.01 mm over a distance of 1016 mm in angularity and parallelism have been reported.

The blocks are torn apart after use and reassembled for another fixture. Some users use a CAD system with a block library to design the fixtures on the screen and keeping the design for future use. Others report using a Polaroid camera to store the design of modular fixtures. It is reported that rebuilding a fixture after a long period of time allows a precision of positioning of 0.08 mm.

Another example: the design of a master jig with additional adapters is one way of dealing with changes in size, numbers and location of features.

The modular fixtures can also reduce the set up times. Using rotary tables allows machining of the four sides of a cube in one set up. Part accuracy improves because the part does not have to be reset. Quick clamping devices, special eccentric cams, slotted bolts and hydraulic or pneumatic clamping can also reduce handling and set up times and are therefore efficient means of improving overall productivity.

3. LEAD TIME REDUCTION

Mr. PM introduced this method by stating that lead time reduction automatically reduces work-in-process. Therefore, their methods may be used specifically to reduce work-in-process.

The capacity planning logic and programs are composed of several stages. The first stage is to examine the feasibility of meeting the delivery date, this is done by computing the slack which indicates if the first operation before or after the current date. A negative slack indicates that delay might occur if normal manufacturing procedures are followed. For normal manufacturing procedures, a generous allowance is made for interoperations such as:

- Queue time (i.e., time spent waiting to be assigned to a machine).
- Pre-operation time (i.e., time for marking out, cleaning, etc.).
- Post operation time (i.e., time for inspection, cooling, etc).
- Wait time (i.e., time spent waiting for transportation).
- Transportation time (i.e., time required for transportation to the next work center).

In case of delay, expediting may be applied and initially allowed time is reduced.

Mr. F: I do not understand why initially such a long time is allowed, even if there is a slack, it does not mean that this time has to be wasted.

Mr. FM: The interoperations are a necessary. You cannot expect that at the instant the an operation is finished, the material handling equipment will stand by it and drive as fast as it may to the next workstation, which is waiting to start processing the next operation. It just cannot be done normally. The question is the allowed planed time.

Mr. PS: Normally we allow a day or two for the interoperation time. This time is normal allowance in any modern capacity planning computer program. For rush cases, expediting is used.

Mr. FM: Expediting is used to ensure that the execution of job orders released to the shop will keep as close as possible to the plan.

Mr. PS: Another scheduling method that is used in cases of anticipated delivery delay is splitting.

Splitting is the simultaneous processing of an operation on several machines. By this means a reduction in operation duration is achieved.

The technical number of splits is determined by the number of similar machines or by the number of tooling sets available.

Mr. F: By splitting lead time is reduced, and thus the work-in-process, but it calls for double setup and thus increase processing cost, I advise that an economic algorithm be used before splitting operations. The economic number of splits is a function of set-up time and operation time.

Mr. PS: Another technique is "Overlap". Overlap is starting the subsequent operation before he preceding one has completed the planned quantity. The results can be a considerable savings in lead time. Fig. 5-1 shows this effect and defines some of the terminology.

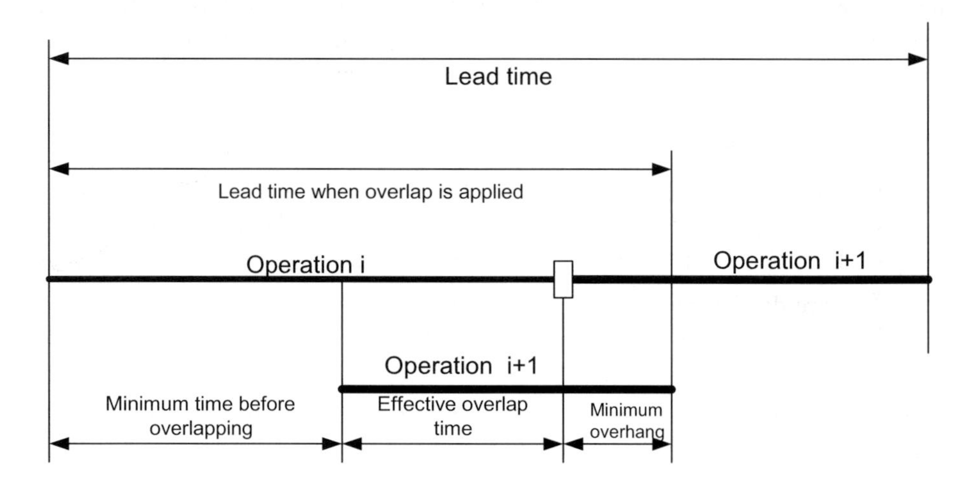

Figure 5-1. Effect and terminology of overlap

Mr. FM: Overlapping for sure reduces lead time, but I am not sure that it is practical as it must be tightly controlled. The overlapping operation may start only when the overlapped operation completes a given quantity and the required interoperation time has elapsed. The setup time of the overlapping operation may start in parallel, so that the operation can begin immediately after items are received. A full time expediting must be employed.

Mr. F: I am not sure that overlapping is always beneficial. The values of minimum time before overlap and the minimum overhang indicates if overlapping can be achieved. If it does not, overlapping is not practical, since the saving in lead time is not balanced by additional coordinating effort.

Mr. CC: Computation can be performed in order to decide whether and how to plan overlapping. If overlapping is worthwhile, the preferable form can be computed by two methods; the overlap is computed forward by using as a constraint the minimum time before overlap, or computing backward using as a constraint the minimum overhang. The method that will give the best results will be used.

Mr. PS: Several times we use a practical rule which is: that if operation i is longer than operation i+1, backward overlap should be applied. When one wishes to overlap three operations in series, with the middle one being short, special treatment is required, since results achieved by normal overlapping are not satisfactory. In such a case it might be advantageous to split the middle operation into a few batches.

The president noted that these methods might be satisfactory for expediting orders that cannot meet delivery date by normal manufacturing procedures. For cost reduction we must employ normal manufacturing procedures. I do not recommend adapting these methods to solve our problem of cost reduction.

3.1 Job release

Mr. C: Before job execution can start, some auxiliary jobs have to be performed. The auxiliary jobs are:

- Fixture design and build
- Tool preparation
- NC program generation
- Material preparation (inventory management and control)
- Material handling (transport)
- Quality control, (preparation of method and tools)
- Set up instructions and Set Up
- Job instruction

In some cases the worker is instructed to perform a job, without having all the auxiliary jobs ready for him at the workstation. He then has to leave the resource and go to look for missing tools, raw material, instructions, etc. While he walks through shop he might stop to chat with fellow workers, to wait in line at the storeroom, to look for an inspector and so on. This is a waste of time, which results with increase lead time, and thus increases work-in-process.

The operator must not leave his post; all auxiliary material should be supplied to him.

Mr. PM: I agree with your comment. We should adopt the KIT method. A kit including all the auxiliary material required should be prepared. Unless the kit is complete, this job should not be released to the operator.

Chapter 6

COMPETITIVE MANAGEMENT

Organization methods

1. INTRODUCTION

The president examined the company performance and noticed a decline in sales and profit. He called a meeting to discuss steps to improve the situation. He opened the meeting by presenting the topic: our company is number one in its field of consumer products and we wish to keep it this way. Now the competitors are starting to reach out. Lets have ideas of steps to be taken. I would like to hear some general remarks from each one of you before discussing the details.

Marketing manager Mr. M proposed to concentrate on developing a marketing strategy that increased product attractiveness to the customers. We should increase our advertisement budget, redesign our product to enable more options, considering shape, ease of operation, increase reliability, durability etc. Practically, I suggest to conduct a market research to establish customers' priority rating.

Sales manager Mr. S: the way to increase sales is by being able to promise early delivery and competitive prices. Therefore I recommend keeping large amount of finished product inventory. We must as well as be prepared to supply any specific requirement and special product option. Therefore, we must keep an inventory buffers between processing stages, and some special raw material available in stock.

Finance manager Mr. F: we must take steps to improve our cash flow to reduce our cost and we keep to a minimum the amount of capital tied down in production. I propose to cut to a minimum our inventory, and direct the monetary saving to other venues that will improve our processing capabilities.

Production manager Mr. PM: One of the cost increasing parameters in production is the frequent change in orders. Any change in order quantity, order specification causes a waste of resources. I proposed to limit the changes to a minimum, or issue the order for processing only after checking and rechecking that the order is exactly what the customer wants. Another parameter is the "rush orders" such orders disturb the planned schedule, causing extra setups and waste of time and material. Such orders should be stopped. We can except such orders once a week when we schedule next week's work load.

Design manager Mr. D: to meet marketing objectives design must be able to meet design specifications and objectives. Today we either get incomplete specifications, or they are being changed here and there. It disturbs our work and causes a waste of design time. Marketing and sales must improve their communication methods with the customers, and issue a design order in final and accepted forms. To improve design and enable to get a speed response I recommend that we purchase modern CAD - computer aided design stations.

Process planner Mr. PP: process planning must meet design specifications. In many cases design does not consider ease of processing. In border cases we do our best to meet the "cost increasing" design features. In extreme cases we contact the designer and negotiate design changes. I recommend design will consider DFA (design for assembly) and DFP (design for processing) in their initial design. To improve design and enable a speedy response I propose to purchase CAPP - computer aided process planning programs.

Quality manager Mr. Q: quality assurance and control activity is to assure that the produced items and products meet the design specifications. In many cases where there are discrepancies a "renovation committee" changes design specifications. The designer is not always aware of the tolerances that he specifies. This case increase in processing cost, in inspection, in managers time. I propose that design be more conscious in specifying features and tolerances. I propose to install a computerized SPC - statistical process control system to improve quality and reduce waste.

Computing Center manager Mr. CC: Production planning and scheduling based its planning on data. The data includes: orders, bill of materials, resources, routing, and inventory and purchasing status. All the above data has to be synchronized otherwise the schedule be unrealistic and cause waste. The main problem is data validation and accuracy. In many cases operation departments fail to report activities, or report incomplete data. In such cases the scheduling and management report are inaccurate and sometimes even misleading. All disciplines must observe reporting procedures accurately and completely. I propose to purchase a new modern computer with higher speed and extended memory in order to speed up and improve our information system.

Foreman Mr. FM: We are supposed to follow production planning schedules. In many cases the schedule is unrealistic, jobs that have been finished is scheduled for the future, material and tools are missing, misleading routing are used for scheduling. Above all are the changes. We start working and a dispatcher comes and changes priorities or quantities. This causes confusion and waste of processing time. I proposed to install a data collection system in order to improve information processing and scheduling.

Inventory manager Mr. I: I feel very bad when I get an issue order and the physical stock is not available while the records indicate that stock is available. Something went wrong, and it is not the physical stock, it is always right. It means that some materials were issued to shop floor without a request slip, or with the wrong quantity or one of many other possible errors or procedures were not followed. Such errors cause idle resources on shop floor, waste and loss of money. We must improve our inventory recording system. I recommend that we recheck our inventory procedures, install data collection systems and increase the data verification method in the computer programs.

Purchasing manager Mr. PR: We purchase what others ask us to but, we do not initiate anything. To make a good buy, which means quality, delivery and cost, payment conditions, etc., several suppliers are contacted with a request for a quotation, so that it takes time, from the request till issue. In many cases we get purchase requests to be delivered yesterday. We ask production planning to give us purchase orders as early as possible in order to allow for ample time to get quotations and arrive at the best deal. I propose to change the procedures to more flexible ones. May we get a yearly or quarterly list of items to purchase? Maybe to move to yearly contract with suppliers? Anyhow thing must be changed in order to serve production better.

The president noted that in spite of the company objective of making profit neither one of the disciplines is working toward this common objective. Each discipline contributes to profit, but indirectly. Not even a single discipline of the manufacturing process considers economics and cost as their primary objective. Each stage optimizes its task to the best of its ability. Each stage, according to its function, has its own primary objectives and criteria of optimization where profit is a secondary objective. Primary objectives must be met. Meeting the secondary objectives is a compromise between several demands, and might reach only, let's say 80% of the optimum.

This leads to the conclusion that management's main objective can never be met. Management is a decision making process, a good, balanced and unbiased decision will be achieved by considering the viewpoints of many disciplines and finding a compromise. Only if each of the disciplines stands up for its own interests will a good balance in the overall operation of the plant be reached.

Therefore I propose that with this in mind we will start our discussions with the objective to arrive at a method that will keep us as the leading company in our field.

Chapter 7

PRODUCT SPECIFICATION

1. RE-EVALUATING PRODUCT SPECIFICATION

Marketing manager Mr. M opened the session with a statement that Marketeers have long assumed that there is a positive association between being "market oriented" and good company performance. For instance, a company operating according to the marketing concept creates profit through customer satisfaction. Therefore we should concentrate on discussing marketing plans.

Mr. F was not happy with that statement and noted that profitability may vary due to several reasons such as the level of investment in R&D, innovative product design, efficient manufacturing. Furthermore there are many ways to define and measure profitability and the effect of market orientation on profit.

Mr. M: The statement that "a marketing concept creates profit through customer satisfaction" is a result of an intensive research by two methods. Two common research methods on the effect of a market orientation on company performance are used: the subjective and the objective. The subjective studies generally incorporate subjective measures of performance as the dependent variables. The term "subjective" means that the company's performance score is derived using a scale with anchors such as "very poor" to "very good, " or "much lower" to "much higher" compared tocompetitors. These can be contrasted with an "objective" measure that would be an actual percentage figure for sales growth or profitability. In many cases these two methods give similar results and confirm the statement made.

Mr. CC pointed out that Mr. M made a general statement that marketing, and I assume that he meant a good marketing plan, is the one responsible for company profitability. He relied on research made in several industries, the question is: does the type of industry, the size, the time fluctuations have effect on the results of the study? Is the term profitability an absolute value or relative to other companies in the same field? I am sure that marketing is important but it is only one of the factors, and I am not the most important one.

Mr. M: May I give a short presentation on marketing functions in the organization.

The perception of marketing has changed dramatically over the years. It started with "the production concept", moved to "the sales concept" and next to what we call today the "the marketing concept". The production concept is that a firm should produce and distribute those products it can produce most efficiently. A typical example is Henry Ford's Model T automobile. He declared that "You can have any color you want as long as it's black", clearly, this approach works as long as a firm has a product that most consumers need, the demand for the product exceeds its supply, and there is little or no competition.

By the 1920s, advances in manufacturing technology had given rise to widespread mass production in the industrialized countries - supply began to exceed demand. The production approach could not be sustained. A new approach was developed. The sales concept holds that just about anything can be sold to customers, whether they want it or not, if the sales approach is aggressive enough. The objective of the sales concept viewpoint is to sell what is available, using all the advertising and personal selling skills one has, with little concern to the customer's post purchase satisfaction. The sales approach also suffered a setback in the early 50s when it became clear that you cannot really sell anything to consumers if they have no need for it. The modern marketing "concept" began to emerge in the early 1950s. A marketing-oriented firm sees itself as in the business of fulfilling consumers' wants and needs, whatever they may be. As such the following functions are considered part of the marketing effort: product, distribution (place), price, promotion (advertising) - what is called the "4 p's" of marketing. That is, the basic tenet of customer satisfaction drives the entire organization and marketing is the driving force behind the actions of all functions.

The president interrupted and reminded those present that the objective of this session was to work as a team and supplement each other, and not to argue about who contributes more to company success. One thing for sure is that a good and saleable product is a promise for profitability.

However, a marketing plan and objectives are part of the business plan which is set by management, and it falls outside the scope of our group. Our group deals with production. The link between production and business is the decisions of product specifications. In this field each discipline must consider the expertise of the others.

Therefore the topic of this session is: we have a line of products let us examine if we can improve them or come out with a new product. In addition to definition and specifications we must make sure that we will be able to introduce it to the market in a very short period of time.

Mr. M: we have a good line of product, however, lately our competitors are copying our product with minor changes, and in several cases we were losing customers. I propose to do two things: one to come with ideas of keeping present customers, and two, adding features to our products.

Mr. F: I buy the idea that we should concentrate on keeping present customers. Checking our budget indicated that most sales and marketing dollars are spent attracting new customers. But getting new customers is about six times more expensive than retaining the ones already in place. This is because of increased advertising and promotional expenses and incremental expenses connected with setting up new accounts. Other expenses include credit searches and operating costs as the firm learns the needs of its new customer, and the customer learns how the firm works.

The key to retaining customers is more than providing "satisfaction" or competing on price. It means an all-out effort to ensure that your customers have an intimate knowledge of your products, and services. This intimacy can be accomplished by implementing targeted, direct marketing campaigns for value added membership programs, aimed at precisely defined market segments. Customer contact is only valuable if it provides customers with value-added products or services. This requires an in-depth understanding of who your customers are and what they want.

Mr. CC noted that using IT (information technology) can be used for this purpose by understanding who our customers are and what products and services they want. We may use CRM (Customer relationship management) software. CRM is defined as a strategy for managing customers and customer relationships. To develop a network of "touch points" with customers that establishes, cultivate and maintain long-lasting relationships.

Mr. C: I agree that it is a good idea to collect and keep customer information database to be used as data analysis tools.

Mr. S noted that such information may be used to form a customer profile that stores knowledge about customers and help to sell morc of a company's products and services more efficiently. By predicting individual customer preferences and needs well enough to be anticipatory and proactive in the delivery of the right message to the right person at the right time via the right media, thus increasing sales and keeping customer loyalty to our company.

Mr. CC continues: The longer a customer stays with a company, the more he or she is worth to the company. The simple truth about long-term customers is that they buy more, take less of the company's time, are less concerned about price, and bring in new customers. Our purpose should be to reduce customer defections: as reduction of as little as five percent can double profits.

The reasons behind customer defection aren't obvious. An intuitive response to defections might focus on customer satisfaction. Ninety percent of customers who defect do so not because they are dissatisfied, but because they have found a tempting alternative. The next largest category of defections due to dissatisfaction relates to the way they have been treated. Customers want to feel important. Dissatisfaction is like an infectious plague. About 75 percent of dissatisfied customers tell at least one other person of their discontent. Only seven percent bother to tell their original service provider. Customer loyalty delivers huge bottom-line business impact because loyal customers spend more, stay longer, cost less to service and refer more new customers.

The president noted that these marketing points are valuable, and will be evaluated by management and the board of directors. In the technical aspects, we wish to create customer satisfaction by offering quality products with needed attributes and at a reasonable price. Consumers do not buy goods or services but rather buy the attributes of those goods or services; hence, success in the marketplace rests on creating products whose attributes match what the market wants and needs. Let's work on this direction.

Mr. S: We have a line of products with different sizes and attributes. These products have been designed in different times and probably by different designers. I am sure that the basic design is the same but probably there are variations in details. As you know the product is made of details. I would like to suggest that we survey the design of all similar products, and try to improve the design by adapting the best detail design.

Mr. D: It is a good idea but I wonder if we will gain any benefit from it. I have excellent designers, and if there are inconsistencies in the design, there was probably a good reason for it. We at the design department are overloaded and I cannot spare any designer to work on past design. May I suggest that Mr. S will conduct a survey to find out the sales amount of each of our products, and present to us a Pareto diagram of product - sales per last year. Such a diagram will show the most saleable product, and the order of priorities in which we should put our improvement efforts.

Mr. S: What Mr. D proposes is a good step. It points to a product that is a leading one in sales within our line of products. However, if this product is being sold in large quantities, it means that its attributes and price meets customer satisfaction. Why should we modify it? I suggested examining a family of parts; where a family means a group of products that serve the same or similar objective, but which comes with different sizes, colors and probably minor variations in attributes.

Mr. D: I agree but before we spend resources in evaluating several product designs, I propose to examine if that variety in product attributes is really worthwhile. Maybe we should reduce the number of variations. Remember that it does not affect only design but also production, and inventory size by storing many items of the same type but different size. It also complicates production planning, process planning, quality control etc.

Mr. Q supports Mr. D's proposal and adds arguments to minimize family size. Each product of the family requires research, preparing documentation, measuring tools etc.

Mr. PP joined the discussion and supported the motion. Each product of the family requires a different process and routine, tools, which increases inventory.

Mr. I joined the discussion and supported the motion. Increasing the number of items in stock makes increasing demands on the warehouse space.

Mr. PM added that increasing the number of products in a family complicates the product structure especially when there are several sub-assemblies some with minor differences. It creates a need for complex capacity planning and scheduling programs, increasing computer load and memory size. In addition it increases the work in process and processing time.

Mr. M was amazed by the collective attack on the line of product. Customers prefer to have a choice of products for the same objective. We know that it increases costs but we think that it is a must. The options increase sales.

The president interrupted and said that this was a management decision. He was ready to hear any comments and suggestions, but it was not a technological decision but rather business decision.

Mr. PM: I do not argue business decisions, but we might ask for design that will keep as many features to be identical in all or several versions of the products. What I mean is to intentionally over design, I know that it might increase direct product cost but we should aim at an overall optimum and not of a specific product optimum design. For example: to use DC power supply of one size and attributes for several products. To use a motor that fit several products, even if some products do not require the power, moment or speed of such a motor.

The president says that he would take it into consideration and proposes to discuss it in our next session: design.

Mr. C: I like the idea that was proposed by Mr. S and adapted by Mr. D of creating family of products. They argue about changing design, reducing number of product options. I propose to use the idea of family of products in order to evaluate attributes. The argument that was made that each design option might have been made by a different designer might hold also by saying that each product specification might have been made at a different time, different market requirements. By knowing the sales quantities of each product in the family we might compare attributes and come up with the attributes of the most saleable product version. Then we can adapt these attributes to all products of that family, and having such improved products will increase sales.

Mr. M: I like this idea. Actually what you propose is to re-evaluate our product attribute, by doing an in-shop benchmarking. Market demands are changing all the time, market is searching for new products and new options. Our data files contain the data required for such research. It can be done in a short period of time, and be in confidence to our competitors.

Mr. I: Could you please elaborate on the method to do it. The objective is clear and recommendable.

Mr. M explained. Suppose we have a family of 5 products, and we evaluate 6 attributes. We construct a table as shown in table 7-1. Sales rating represent the quantity sold of each product. 100% is assigned to the product with the higher sales, (in this case product 1) and for the others the rating is assigned as a sale relative to the product with the highest sales. The column 'attribute' lists the six attributes. In the row of each attribute an "X" marks that this attribute is available in the product column, a "-" means that this product does not posses that attribute.

Examining table 7-1 we may write the following equations:

Attributes A+B+D+E = 100% sales rating - product 1 column [1]
Attributes A+C+E = 55% sales rating - product 2 column [2]
Attributes B+F = 30% sales rating - product 3 column [3]
Attributes A+B+C+F = 70% sales rating - product 4 column [4]
Attributes B+C+D+E = 60% sales rating - product 5column [5]

By these 5 equations we can compute the sales rating of each attribute:

Equation [1] indicates that sales rating of C and F = 0
Equation [3] indicates that B+0 = 30%; hence B = 30%
Equation [4] indicates that A+30+0+0 = 70%; hence A = 40%
Equation [2] indicates that 40+0+E = 55%; hence E = 15%
Equation [5] indicates that 30+0+D+15 = 60; hence D = 15%

Table 7-1. In-house Benchmarking

Attribute	Product 1	Product 2	Product 3	Product 4	Product 5
Sales rating	100%	55%	30%	70%	60%
A	X	X	-	X	-
B	X	-	X	X	X
C	-	X	-	X	X
D	X	-	-	-	X
E	X	X	-	-	X
F	-	-	X	X	-

The conclusions that may be drawn from the correlations between sales and attributes that attributes C and F have no effect at all and it is just a waste of money to incorporate them in the products.

Product 4 is an extension of product 3 and we should consider abandoning product 3.

Mr. S was impressed by the easy and simple method of evaluating the effect of attributes on sales. However, he questioned the term “sales rating”, there might be two meaning; one in cash (dollar) and the second in quantity. I am sure that the figures will be different in ease case. Product 3 is actually a simplified version of product 4, therefore its price is about 40% of the price of product 4, and therefore the quantity sold is three times that of the quantity sold of product 4. In such case the revenue to the company in keeping product 3 on the line is larger than that of product 4, and we might keep it as well.

Mr. F joined the discussion and asked if we had data on the cost of each attributes. The question of rating should not be that of sales volume or income but of the profit we have by selling each product.

Mr. D noted that marking an “X” if the attribute is present in a product, is an oversimplified assumption. There are many ways to incorporate an attribute in design. For example: a time indicator attribute on the product may be in digital or analog form, to show only the hour and minutes or to have seconds and date as well, the size of the display, to have light on the background etc. In all such cases an “X” will be indicated on this attributes.

I like your method of evaluation, but like to add that the attributes should be indicated by value and not by just an "X".

Mr. CC responded that we keep data on the cost of each item that is part on the product structure, produced or purchased. But we do not have the cost of assembly for each individual item in a sub-assembly or the whole product assembly. The data collection keeps the total assembly time (cost).

Mr. M agreed with all the comments made. These comments should be taken into consideration. In my presentation I wanted to point out that we should re-evaluate our product design once in a while. Market demands are changing all the time, market is searching for new products and new options. I gave just a very simplified example of how the re-evaluation can be done. For practical evaluation our criterion of operation should be profit made by each product. The simple table 7-1 must be expanded to include all family products, and a long list of attributes that should be considered. Naturally more sophisticated mathematical techniques should be used in evaluation and decision making.

Mr. C: That sounds an excellent method. But why should we use it only for evaluation in-house products. Our goal is to be number one in the market, so why not use it for evaluation of the competitor's products as well. Consumers do not buy goods or services but rather buy the attributes of those goods; hence, success in the marketplace rests on creating products whose attributes match what the market wants and needs. Knowledge of the market value that is attached to each of the most important attributes of a technology-based product is important information for managers. Many businesses are built on products that have a single outstanding characteristic that none of the competing products can match, while satisfying minimal standards in other characteristics.

Mr. FM supports the notion to do an all over evaluation, but he is concerned about getting data for it. We know what is going on in our company, we can see the competitor's publications, and product cost. But we need more intimate data, which the competitors, as well us, keep as secrets. How can we acquire those data?

Mr. CC volunteered to answer this question. A multitude of tools, techniques, products, and services have lowered the costs and vastly improved access to information for competitive intelligence researchers - immensely simplifying data gathering.

The World Wide Web changed the situation. Among the Internet's greatest contributions to competitive intelligence research is the window it opens onto business relationships. The Internet can expose a variety of relationships that may not be widely publicized. Moreover, either party may not even approve the information uncovered for broadcast. The hyperlinks of the Internet can be used as direct evidence of official and unofficial relationships, with real value in the links pointing to a particular Web site. The Internet also excels at providing swift access to critical news about your rivals. While the Internet offers effortless access to almost limitless information, the accuracy of the information must be suspected.

Some fairly straightforward Internet searches can reveal a multitude of relationships - and hopefully more details about the alliance. One may uncover a rival's client list, a rival's supplier, or details about the competitor revealed in a success story. Such searches don't specify the direction of the relationship: the same search may uncover a rival's clientele, or the companies to whom your rival is a client.

Along with opening a window on business relationships, the Internet excels at providing swift access to critical news about your rivals. Certainly, electronic clipping services existed before the Web. However, the electronic highway has expanded the variety, simplified access, and lowered the cost of these services. Beyond just watching for news stories out in the print world, attentive services will monitor changes in Web pages you specify, seek new filing or patents, or continuously monitor the Web through a search engine.

Some free Web sentinels are providing an integrated assortment of useful data for company research. Offering valuable Web-based monitoring of public companies, this will monitor up to ten U.S. public companies. When your selected companies submit receive patents or trademark approvals, post jobs, release news stories, register internet domains, or are mentioned in several investor focused discussion groups-you receive an Email alert. Other possibilities include silently watching your specified pages, and then notifying you via email when changes take place. A setup can monitor a competitor's home page (or any page you specify), signaling with important hints when your rival posts news stories, jobs, executive speeches, new products, and more. Other possibilities: monitor the Web site of a rival's hometown newspaper, specifying the rival's name. When the electrons hit the wires with a news story, you receive an email alert.

Free simple search capabilities can be accomplished on four well-known search engines: AltaVista, Excite, Infoseek, and Lycos. Informants will also monitor Web pages you specify and notify you when changes take place. The possibilities are limited only by your imagination: patrol for executive speeches, distributors, new locations, trade show exhibits, research papers presented at conferences, whatever strikes your fancy.

Discussion groups on the Internet are serving as Internet era watering holes - offering facts along with gossip, rumor, and innuendo on a wide array of topics including investment-related information. The discussion groups can provide hints and tips from industry and company experts, or they can provide off-the-wall comments from uninformed eccentrics.

Over the past few years, businesses, and inventors have benefited greatly from free public patents, and trademarks on the Internet. Another large cache of company-specific information is just now finding its way onto the Web-public records from state, county, city agencies, and federal courts.

For the Internet, the data needs to be verified. Verify the data with another resource. Verify the author. Verify the date of the information. Verify the domain's owner. Even if you find data at a company's home site, it could be misinformation, designed to mislead-something for which the company may have to plead forgiveness in front of television cameras at a future press conference.

Mr. S: What we intended to accomplish was to improve our products by adapting features that are in the products of our competitors, and the customer probably wants. May I suggest that in addition we improve our products by incorporating attributes that are not available today on the market?

Mr. M was surprised by this proposal and asks for clarifications.

Mr. S continued: Each product is designed to perform a specific main task, which is closely related to the core function of the product. As an example consider a scale which has a core function of the ability to measure weights. Or telephone which has as a core function, the ability to receive and transmit voice and data. The product can be extended by combining traditionally separate products and services, responding to demands for new services and embedding new services into traditional products. As an example with industrial scale (which is available today) add an on-line reading, controlling the packaging to a specific weight, etc. The telephone may be extended to transmit pictures, facsimile, wakeup call etc.

The president was very pleased by this augmentation and asked to treat it seriously. Such an approach extends the scope of our products and will keep us number one in our field.

2. PRODUCT SPECIFICATION METHODS

Mr. M opened this session by stating that market demands are: short time to market; product diversity and options; quality products. Customer satisfaction and, naturally, competitive price are basically controlled by product specification; product design; and process planning. These three functions control the minimum cost of a product and the lead time to market. However, product specification as well as product design are innovative tasks and depend on designer creativity.

Product specification has to seduce the customer, with its options and appearance. To arrive at such specifications many disciplines of the manufacturing cycle should be involved. A comparison to similar products, produced all over the world, must be made (benchmarking, One of a Kind, World Class Manufacturing, etc.).

Well-balanced and unbiased decisions will be achieved by considering the view points of all disciplines and finding a compromise. The need for such a compromise is commonly accepted. The problem is: How to arrive at such a compromise. The easy solution, and not an imaginative one, is to set committees and group discussions.

Mr. D noted that he already serves in several committees and it interferes with his routing jobs. From his experience such a committee is just a waste of time. To save time he proposes that let marketing do its work and propose product specifications, circulate this specifications, and let any department express their notes.

Mr. PM noted that a committee is a good place to meet the managers of the company in a relaxed atmosphere, and exchange ideas as we are doing now. However, a product specification working committee is a different story. Market demands for short time to market calls for frequent and long meetings that will interfere with our daily work. I support Mr. D to let marketing prepare a proposal and we, at our leisure, will read it and prepare notes and comments.

Mr. M: your proposed method was tried many times in many organizations, the result was that it took several months to arrive at agreeable specifications. Some comments and request made by design was completely denied by production, therefore the arguments were lasting on end. Getting all to one meeting that might take a week or so, but it is the fastest way to get product specifications at a short period of time.

Mr. C supported marketing to have a committee as each discipline have their own jargon and way of specifying attributes. Management and marketing may be convinced that they have specified an attribute clearly and completely, and properly, while design might not agree or understand what they mean. Phrases such as ‘nice’, ‘easy to operate’, ‘light weight’ are general verbal terms, but are not engineering terms. Management and marketing are not always aware of the meaning of those specifications on production costs and lead time.

It is a natural tendency for the one who specifies product characteristics to aim for the best, and rightfully so. However, he is not always aware of the costs and manufacturing implications. In many cases, reducing the specified values by as little as 5% may result in cost reduction of more than 60%. We assume that the product specifier may change the specifications once he is aware of this effect.

Mr. PP agreed and said: let me demonstrate an example from a bathroom scale specifications. The accuracy of the scale has to be specified in engineering terms and not in management terms. Management might have specified the accuracy as of +/- 1 gram. It can be done. However, it calls for more costly accurate sensors and support circuits. If the accuracy will be specified as +/- 100 grams, a simple sensor and a low cost product will be result. To prevent such misunderstanding it is important to have all disciplines working as a committee.

Mr. C: It is a good example, and probably there are many more examples. This shows that achieving a level of feasible accuracy by mail will take couple of days, while within a committee this can be resolved in minutes. However, what is the alternative?

May I propose that whenever a project of introducing a new or improved product, a bulletin board is set up that describes in general terms the objectives of that product, some of the design constraints, and ask each discipline to prepare a list of attribute proposals. Following the bathroom scale example design will prepare a list of recommended practical accuracies and thus the specifier would be able to make a sound decision.

Mr. S noted that by considering such a recommended list without the production lead time and cost of each accuracy value would not be of real assistance. How can the specifier reach a decision without having such data? He must have data on the consequences of each decision, not only data on cost and lead time, but also on the effect on sales and customer preferences. It brings us back to the alternative of a committee meeting were all disciplines involved will be present, and discuss in real time the pros and cons of each alternative.

Mr. PP noted that decision on an attribute such as presented by the example, is not a matter of pulling data out of a hat. There are many design and process planning alternatives, a decision of which one to select calls for computations, creating alternatives etc. I do not believe that a committee can make sound decisions in one meeting. Probably an issue is raised, ideas and proposal will be presented, and then a conclusion to end the meeting and set another meeting date within couple of days or weeks, letting each discipline evaluate each proposal, compute, consider and reach a decision and arguments for his standing on the matter. Then, it might be that in one additional meeting a sound decision will be reached.

It should be remembered that there will be several issues to discuss, and each one might take several weeks.

Mr. F: It seems that there are arguments against product specification by a committee, but there is not any positive proposal. May be we should form a committee to decide how to act.

The competitive market of today imposes requirements such as: Short time to market; Product diversity and options; Quality products; Customer satisfaction and customer seductiveness and naturally competitive price. Product specification, product design and process planning control most of these demands. Introducing a new or improved product, by wasting time in committee discussions, will put us out of business. It is a race and we are not the only ones that consider new lines of products, and the company that will be the first to introduce the product will control the market.

Mr. CC: I read about a method that might be our solution. It is a computerized committee, it functions as a normal committee but the computer acts as the representative of each discipline. The results are that decisions can be made within days.

Mr. M said that he does not believe in such magic systems, and proposed to stick to the good old methods.

The president interrupted and said that it is worthwhile to understand what a computerized committee is. He asks Mr. CC to give a presentation on that method.

Mr. CC explained: The proposed system is activated by a computerized product specification workstation as shown in Fig. 7-1 and dialog sessions between the user and the computer. The **MASTER PRODUCT DESIGN** is the main program that controls and navigates the session. The general data of the session, such as title, name of user etc. are being recorded in the **specific product design** files. The session proceeds by presenting inquiries to the user, asking him to specify his intentions and wishes. Messages, drawing the attention of the user will be presented whenever the respond to the inquiry will presumably be out of standards or out of reasonable values.

To determine if and when to post a message a **Technical Data file** is used. It contains data relevant to the master product, i.e. it is a general data for a family of products. With the guidelines of the master product design session the customer converts the master product to a specific product that meets his needs.

This system calls the disciplines of marketing, sales, customer relations, inventory control, purchasing, bookkeeping, shipping, packaging, finance etc, by a short name group B. **Group B Note File** contains any notes, comments, attribute requests, made by members of this group. Good decisions will be achieved by considering the view points of all notes posted in this file, and will attempt to find an optimal compromise. Notes of extended product as was discussed in the previous sections are kept in this file.

A **Check list** serves as a reminder by drawing the user's attention to many topics, such as: Ease of operation; Durability (product life); Reliability (low maintenance); Efficiency (low operation cost); Safety; Ease of maintenance; Noise level; Weight; Floor space; Aesthetics; etc. Most of such specifications are of secondary importance, but they enrich the product and make it compatible with other products of the same field. The user may either specify that the topic is of no relevance to the specific product, or he may ask for additional requirements to the product specifications.

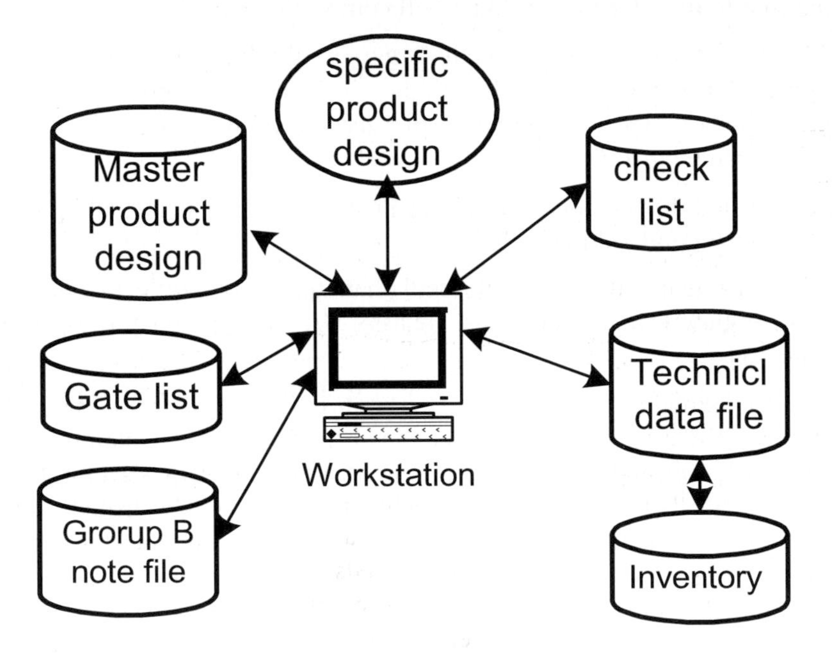

Figure 7-1. Product specification workstation

The **Gate list** is the project manager's tool to control and follow up on the project. It is composed of several gates, each one represents a major milestone in the project development. Each gate is then divided into internal milestones. The objective of the gate system is not to design or make decisions, but rather to control the advancement of the project and make sure that nothing was forgotten or neglected during the course of the development.

Mr. M is skeptic and wonders how a computer can be so smart, and know all the answers?

Mr. CC explained that computer does not know anything unless data is fed to it. The main effort in constructing this method is to collect engineering technical data, and market data. It is a huge job, but it is worthwhile. Study proved that within a few days an optimum product definition that gets the approval of all disciplines has been specified. The proposed system is not a computerized system; it is a technological system, assisted by computers. One cannot have a standard, off-the-shelf product like this. It is unique to each plant. The supporting computer program is very simple to write.

Mr. D: What does the computer know about our products? There must be a correlation between plant techniques, resources, manufacturing know-how and product definition and processing.

Mr. CC: As I said the system is not an off-the-shelf product, it supplies the framework, but the technical data must come from the company. That means that company data have to be taught and captured by the system and kept in the technical data file.

Mr. F: I do not understand why we have to spend a lot of money in order to teach a computer what we already know. It is our data and it is kept in our files or engineers. We do not need a computer to teach us what we know already. Product definition is an innovation process and, as such, must leave freedom and judgment to the individual that performs this task. A computer does not possess innovation capability. It might be of help in production planning, but not in product developments.

Mr. CC explained that the objective of the product definition module is to assist, and not to make decision, merely to guide the product specifier in defining a product that will meet all product objectives. To draw his attention to the effect of any decision that increases product cost and lead time. In some cases the system will propose alternatives. However, the final decision is left to the user. The role of the computer is secondary. The backbone of the system is technology and not computers.

Mr. D noted that all we hear is a sales talk. Can you convince us by letting us know more details about the system? What do you mean by technical data file, what type of data is stored in its file, how accurate are they? Are they theoretical or practical data? Please elaborate.

Mr. CC explained. The master product design is not a universal system; it is aimed to assist a specific plant that is in a specific line of business. The gate module might assist in designing a new product, but normally we assume that the plant has been in business for quite some time and has a line of product. The steps of constructing the system are as follows:

1. The first step is installing the master product design system is to study the existing product. A block diagram of each product will be

drawn. The block diagram should not be a detailed one, but rather present the main module of the product. The block diagram should indicate the objective of each module, and the interrelationship of the different modules.

2. The block diagrams of different products are compared in search for family of products. A family of products is defined as a group of products having the same or similar block diagram. The block diagram of the individual product in the family need not have exactly the same block diagram, but such, that the same blocks appear in most products diagrams.
3. The family block diagram is declared as a master block diagram, and the available products that are included in the family are listed. For example, the master block diagram of weighing scale products will be:

   Platform, load cell, display, options.

4. The objectives of all products in the family are the same, but there can be differences in design among the individual products. The reasons for the differences are studied in order to specify the products. Such a study might reveal that the purpose of the products differs or that the products are used in different environment and require different characteristics. The study of the characteristics indicates how to specify a product. The results of such study are entered in the technical data file. The purpose of this file is to provide assistance and guidance in forming the dialogue session, and to supply data for messages during the session.

Considering the above study a simplified sample of technical data file is as shown in Table 7-2. The values are the range of the different products that the company made in the past.

Table 7-2. Example of Technical Data File for scales

Scale Types	Capacity	Accuracy	Repeat-ability	Linearity	Read-ability	Stabiliza-tion time
Kitchen	10Kg	0.5g	±1g	±2g	2g	2 sec
Bath room	150Kg	100g	±5g	±10g	100g	4 sec
Retail	30Kg	1g	±2g	±2g	1g	2 sec
Industrial	250Kg	25g	±10g	±5g	5g	1 sec
Analytical /Lab	150 g	0.001g	±0.0015g	±0.003g	0.001g	2.5 sec

Mr. S was not impressed by this table and noted that such data is already available in sales bulletins and catalogs, what was wrong with that, why should we change it?

Mr. CC agrees with Mr. S that the data is already available in the company files; therefore arranging them in the form of the technical data file will be easy, fast and not expensive. Supplement data may be acquired from competitors, journals, etc., and added to the file.

Mr. M was not impressed: OK, we have the technical data file, what do we do with it?

Mr. CC explained. This file is used to form the product specification dialogue. The user's final responses to the dialogue are stored in the specific product design file and are regarded as the customer specifications for the product. Continuing with the example of the weighting scale products, the session to define product specifications proceeds as follows:

Computer display: State your desired product
User: Help
Computer : choose one of the following: Kitchen scale
Bath room scale
Retail scale
Industrial scale
Analytical/Lab
User: Retailer weighing scale
Computer: State the required maximum weight
User: 60 Kg
Computer: Are you sure? The normal capacity of a retailer scale is 30Kg
User: OK, make it 30 Kg
Computer: State the Stabilization time
User: Help; what is Stabilization time
Computer: The time that the display is steady
User: I want it to be 0.5 seconds
Computer: Are you sure? The normal settling time is 2 seconds.
User : It is too long for me. Make it 1 second.
Computer: OK, it will be 1 second.
Computer: Select d splay type
User : Help
Computer: Choose one of the following: Digital
Lever pointer
Hand weights
Set limits
User : Digital
And so on.

Mr. M was impressed by the speed of getting response from the computer and the ease by which the user can reach realistic decisions. The system is heavily dependent on forming the appropriate program. Probably the marketing and sales questioner may be used to prepare the computer dialog program.

Mr. S was not so impressed, and noted that it is not a new idea; many companies nowadays let customers specify a product by choosing models from a catalog. I do not see the difference between such system and the proposed one. In both cases the customer assembles a product from modules.

Mr. M does not agree, modules are decided by the company, and thus the number of product combinations is limited. While in the proposed system there may be unlimited number of products combination. It almost reminds one of the new trend of mass customization. I like this ideas, it is similar to the one we had before by evaluating our product attributes (section 1) but this one is more versatile, computerized and can serve some other purposes.

Mr. D joined that opinion, and stated that design probably can use the technical data file, inserting engineering data, and it will become a system. It will speed up design as well as specifications.

The president summed up this session by reminding that the objective of the symposium was to increase our competitiveness. This session was devoted to evaluate product specifications in order to come up with cost reduction methods and being competitive. The computerized system might be a good idea but it probably takes a long time to program it. Doing the attribute evaluation and comparison, is not efficient but it can be done at an earlier time.

Therefore I ask:

1. Mr. CC will evaluate the difficulties of developing the computerized system,
2. Marketing and sales to evaluate the attributes in our products, and to recommend unifications, and new products.

I expect to receive their evaluation within 20 days, and then a decision will be taken.

Chapter 8

PRODUCT DESIGN

1. INTRODUCTION

The president opened this session reminding those present that the objective is to work as a team and supplement each other, and not to argue about who contributes more to company success. Our group deals with production technology. The topic of this session is cost reduction by means of product design.

Mr. PM wishes to recall the previous session (chapter seven) where he, and others, pointed that increasing the number of products in a family complicates the product structure, especially when there are several sub-assemblies, some with minor differences. It creates a need for complex capacity planning and scheduling programs, increasing computer speed and memory size. In addition it increases the work in process and processing time. If, as marketing claims, that option increases sales, then we might ask for design that will keep as many features (components and items) to be identical in all or several versions of the products. What I mean is to intentionally over design, I know that it might increase direct product cost but we should aim at an overall optimum and not of a specific product optimum design. For example: to use DC power supply of one size and attributes for several products. To use a motor that fits several products, even if some products do not require the power, moment or speed of such motor.

Mr. D: This request is reasonable and can be done. Let me explain what design is: The purpose of design is to transform the objective into detail set of engineering ideas, concepts and specifications. Engineering design theories are employed, the objective is translated into engineering specifications and the engineering task is defined. Thus it is a process of innovation. Many ideas and concepts will be formulated and analyzed, and the best conceptual solution will be determined. This conceptual solution will define additional, lower level (detail design) engineering tasks until the last detail of the design is decided upon.

The optimization criteria for the decisions made in design are for the most part engineering considerations: weight, size, stability, durability, ease of operation, ease of maintenance, noise level, cost, and so on. Some of the criteria usually conflict with each other, and thus the decision will often be a compromise. However, the designer's primary criterion in making a decision is to meet the product objectives. This is the designer's most important responsibility, since errors in production are not as critical as errors in design. To be on the safe side, the designer will tend to incorporate as many safety factors as possible.

In case that the design objective is to design a product with several versions and options, design will keep the number of features to a minimum. However, if after a few years a request to add a version is made it is most probable that the number of features will increase.

Mr. C: It does not make sense; probably the designer will study the original design and try to meet the new objectives with minimum changes.

Mr. D does not agree. Technology might have changed during the time that the original design was made and the present day. Changing the design according to present time technology will probably improve and reduce product cost. No one would modify the design of the previous products. Moreover, the designer that is assigned to design the additional version is not the one that designed original version. Each designer has is own ideas and ego, and he wishes to express himself. Thus variations might occur.

Mr. PP: Designers are highly individualistic. No two designers will come out with the same design. It is very rare that designers are able to describe how and why they have chosen a particular solution. Nevertheless, good designers try to convey this subjective process objectively by following a certain pattern. This pattern represents a general problem-solving technique rather than a solution to a particular problem.

Mr. D laughed and said look who is talking. I may say exactly the same on process planners. It is true: There is no single solution to a design problem, but rather a variety of possible solutions which surround a broad optimum. The solutions can come from different fields of engineering and apply to different concepts. The driving power in a machine, for example, can be electric, hydraulic, pneumatic, mechanical transmission, or an internal combustion engine. Among all of these possible solutions, there is an optimum one that is determined by the criteria of the designer.

Mr. FM: Mr. PM asks the designer to over design in order to reduce the number of subassemblies or items and thus reduce inventory and cost. Actually in many cases the designer over design by setting high values of safety factors, in many cases it is just a waste of material and processing time. It will be more productive to aim in the direction of reducing number of items, increasing batch quantities.

Mr. D explained this tendency of designers; the designer faces the problem of predicting the performance of the design. There are many uncertainties, since not every characteristic can be computed on the basis of theoretical scientific knowledge or backed by practical experience. Nevertheless, the designer must make decisions, take responsibility, and hope to achieve an acceptable solution.

Engineering is the application of available scientific and empirical knowledge in the creation of an appliance or machine intended to perform a given task. This given task is defined by either a customer or management, usually in a short qualitative statement. The designer is a problem solver who applies such fields as physics, mathematics, hydraulics, pneumatics, electronics, metallurgy, and strength of materials, dynamics, magnetic, and acoustics in order to find the solution, namely, the new product.

The designer is bound by constraint conditions that arise from physical laws, the limits of available resources, the time factor, company procedures, government regulations, and morality.

The design is iterative. The designer continually reexamines his previous decisions in the light of new information gleaned as the design progresses. New, random ideas and concepts are applied until satisfactory results are obtained.

Mr. PM noted that there are two distinct phases in design work:

1. *Design of the basic concept of the solution.* In this phase designers employ their creativity. They are able to let their imaginations run free and come up with any wild idea. The more extravagant the idea, the better the designer is.
2. *Decision and solution specification.* In this phase designers employ patience and technical know-how. They are constrained by rules, procedures, mathematical equations, and standard communication techniques.

What Mr. D explained refers to phase one. Phase two is straightforward; there are precise equations to compute the variables. The safety factors that were referred to are established in phase two, and probably can be reduced without interfering with the functionality of the product.

It is not easy to switch one's state of mind and work on both phases; however, it is essential. Frequently, the difference between good and bad design resides in a lack of attention to details rather than in the basic concept. Details are frequently left to junior designers or draftsmen who are not fully aware of the problem and its solution. However, the process planner and the production engineers are obliged to accept these drawings without question.

Mr. F agreed that we should look for overall efficiency and cost reduction, and not to design for individual item efficiency. Naturally, it should be checked to assure that the increase in cost of the individual item is compensated by cost reduction in other phases of production. The designer must exercise some common sense in his decisions. A mathematical minimum cost is not a practical optimum; it depends on the optimum curve. In many cases the cost-design curve is very low which means that the optimum indicates a range and not a unique point. Within such range the designer may choose the appropriate value.

Mr. D agrees and asks if Mr. F can set some rules or limits for cases to prefer previous design over a new design. Furthermore, even with such rules there are cases in which the previous design must be changed. For example: the extended product, as was discussed in chapter 7, calls for a new attribute to be incorporated in the old design. The added attributes might be from a different field of technology, suppose it needs a loudspeaker, the loudspeaker needs mounting brackets, which calls for design changes.

Mr. F was astonished and unprepared to answer this question, but as the request rose he had no choice and said off hand that for minor variations which are within 5 to 8% he recommend to prefer the previous design.

The president interrupted and recommended to use these values for the time being, and Mr. F would calculate a more reliable figure.

Mr. S: What I propose in our discussion on product specifications may hold for product design as well. I proposed that as we have a line of products with different sizes and attributes. These products have been designed in different times and probably by different designers. I am sure that the basic design is the same but probably there are variations in details. As you know the product is made of details. I would like to suggest that we survey the design of similar products, and try to improve the design by adapting the best detail design.

Mr. C: We are talking about two separate issues; design of the present products, and design of new versions or products. The products that are on the market today probably cannot be changed, and thus the design with all its variations must be left alone unchanged. We should concentrate on new designs.

Mr. S did not agree with that statement; We can always come with a new model of the same product, while preserving its function and attribute the internal shape and design might be changed without affecting sales. The changes will unify the design details by using a master design, incorporate the best details that were in any one of the product variations.

Mr. F supported the idea of reviewing all product designs and establishing master product designs and using standard design and standard features and components.

Mr. D rejected the idea. Such a review can be made only by the designer, and they are very busy; beside it will take a long time and be expensive. It probably will not be cost effective.

Mr. F insisted that it might be cost effective and the justification is: in introducing a new part there is the design time, detail drafting time, prototyping, testing, and documentation and certainly drawing maintenance. When the new part design hits manufacturing many things happen. There is advance manufacturing engineering from a central location and possibly at remote plant location. There is tool design. Tools have to be either made or bought. Time study is involved. Production control has to schedule the part, cost accounting is involved, data processing, purchasing, quality control, N/C programming are all affected – we could go on and on. It is expensive to support new parts. But we can carefully examine the active parts of the company, and create a family of products and parts, and make them company standards. When a new part is required, before rushing to design, we can retrieve and compare the available parts to decide if they can be used. The cost of introducing a new part is about $3000. By using standard parts, experiments show that at least 5% of new required parts can be avoided. In a modest size company $250,000 to $500,000 can be saved, by not designing new parts.

Mr. PP: It is not enough to compare existing designs and selecting the best one as a master design. The master design has to be evaluated for improvements. Designers have a tendency to over-design by using a generous safety factor. For mechanical components it is customary to use a factor from 4 to 40 depending on the designer's temper. The safety factor can be obtained by designer intuition or by considering each determining factor separately and multiplying each other such as "Engineering in the Manufacturing Process Safety"; Thus safety factor - SF can be calculated by:

$SF = FA\,(L1 * L2 * L3 * S1 * S2 * S3 * S4 * S5)$

FA = working safety factor
= 1.0 if the failure is not serious
= 1.2 if the failure is serious
= 1.4 if the failure is very serious

L1 = uncertainty in magnitude of load applied

L2 = nature of applied load
L3 = uncertainty in load distribution
S1 = variation in material properties
S2 = manufacturing effects
S3 = environmental effects
S4 = stress concentration
S5 = design assumptions.

The following values are recommended for L1, L3, S1, S2, S3, and S5

Very good	1.1
Good	1.3
Fair	1.5
Poor	1.6

The following values are recommended for L2:

Light shock load	1.2
Medium shock	1.5
Heavy shock	2-3
Impact	>3

For example one uses for modest cases a safety factor of 30

SF = 1.2* (1.5*1.5*1.5*1.5*1.5*1.5*1.5*1.5) = 30

It means that by assigning the size of a mechanical component the size will be 30 times larger than that the strength computation indicates. It is just a waste of material and processing.

Mr. D was not happy with this demonstration, and said that the designer must have some margins. Judging by the end design one may reach the conclusion of waste of material, but in many cases the controlling parameter are not always the strength, for example: the strength calculation call for a size of Ø0.2 mm, but the bending and deflection are the controlling factors, and they call for Ø2.0 mm, thus you cannot say that a safety factor of 10 was used.

Mr. FM: I agree that a safety factor must be used. But it keeps a large margin which can be used for unifications of items in several versions of a product. Use the same item; in one product you might have a safety factor of 20 while in another a safety factor of only 10, which might be good enough.

Mr. PM: An excessive safety factor gives the designer peace of mind and security; however, it can also give rise to severe penalties in weight, size, and cost. Another approach to the selection of the safety factor is illustrated in Fig. 8-1. Both the load capacity (strength) and the actual load (stress) are not fixed values, but due to the nature of the design have a certain distribution around a mean value. The specific shape of the distribution curve depends on the particular problem. The safety factor is defined as the ratio of the mean load capacity to the mean actual load. The overlapping area of the distribution curves indicates the probability of failure. The designer can choose any desired reliability value by using statistical theory to compute the corresponding safety factor.

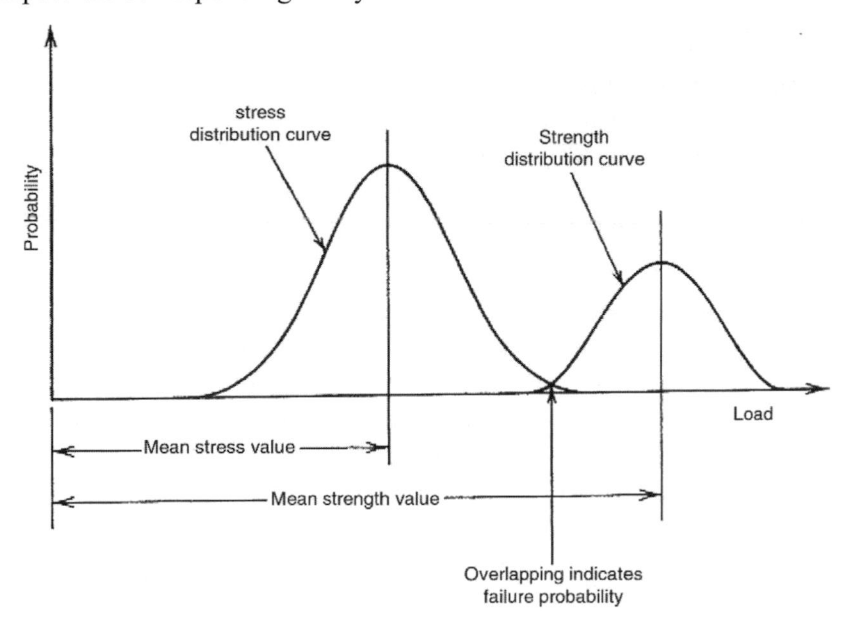

Figure 8-1. The relationship between the distribution of strength and stress

Mr. PP: The safety factor covers cases of strength failures. There is another potential of error in manufacturing that can either affect production performance, product life, and product assembly or have no significant effect on the product at all. In manufacturing, it is impossible to make each dimension and characteristic agree exactly with one specific value. Every element will deviate from the theoretical dimension to some extent. In many cases even gross deviation from component geometry and characteristic can exist with no significant effect on product performance. On the other hand, in some cases a microscopic deviation can have a catastrophic effect.

To ensure against failure due to human error in manufacturing, the designer specifies the permissible deviation, that is, the acceptable range of values. In other words, the designer specifies a tolerance. In mechanical parts there are three types of dimensional characteristics which need to be controlled by tolerances: size, shape, and location.

Tight tolerances afford the designer peace of mind and security; however, they also raise the cost of the product. In many cases tight tolerance of an open surface is assigned which does not make sense. But my job is to set a process that will meet drawing specifications. I am sure that with tighter control on the designer tolerance setting will reduce processing time and cost.

Mr. PM: I agree with Mr. PP that in many cases we detect too tight tolerances, and it is a constant conflict between the manufacturing department and the designer. The increase in cost with tighter tolerances is due to the following:

- More machining operations.
- More rejected items.
- More expensive gauges.
- More expensive quality control.
- More expensive machining tools and fixtures.
- More set-up time for processing.
- Higher wages (skilled labor is required).

In engineering handbook recommendations where tolerances and safety factors are given, these recommended values usually include safety margins, and there is no need for the designer to add additional factors. Only in special cases when the designer computes design values based on special equations he should add safety factors. Adopting such a procedure might reduce processing cost and time by about 25% and most of all eliminate the conflict between manufacturing department and the designer.

Mr. Q: Beside the savings that can be accomplished by controlling the safety factor and tolerances, there are also savings that can be achieved by the design itself. For quality control and process planning the design as presented by the drawings is the yardstick for performance. It must be met, and there is no consideration of why any specified feature was asked for. In many cases an item is rejected because it did not meet drawing specifications. When we ask the designer to change the drawing, in order to salvage the part, many times the designer does it willingly. When asked why he did not do it originally he responded that the shape of dimension is of no importance to the performance of the product, but he had to specify a specific data and so he did.

Mr. D explained this phenomenon: Product and part design must meet their primary objectives and represent a compromise between several secondary objectives. Any designer or plant might attach a different degree of importance to any single secondary objective, thus different designers will end up with different designs. In a way, design is a work of art that carries the personality of the designer. There is nothing wrong in this, since there are many ways to achieve a goal. In the design of any part, there are usually a few functional dimensions and shapes. The other dimensions and shapes are fillers. However, the engineering drawing must include and specify all part details. It is estimated that about 70% of the details in a drawing are fillers, and only 30% are functional.

The functional details can be divided into three categories:

1. the detail is constrained by the function that must be performed;
2. the constraint is due to purchased items (for example; bearings) that are assembled in the product;
3. the constraint is due to a decision made previously on another component (number and size of holes in a cover plate).

The last two categories can be changed by selecting a different standard item (the variety is large) or by revision of previous decisions.

Thus, from a performance standpoint, the designer has a great degree of freedom. This freedom is utilized in design for economical manufacture. It is important for the designer to be aware of the various design features, which can affect manufacturing costs.

Some of the factors that can increase the ease with which a product can be assembled are:

1. reduction in the number of parts;
2. provision of a suitable base component;
3. assembly in layer fashion from above along a vertical axis;
4. elimination of lengthy fastening techniques; and
5. design for ease of locating or aligning by providing chamfers, pilots, and so on.

Mr. PP: Thus, from a performance standpoint, the designer has a great degree of freedom. This freedom should be utilized in design for economical manufacture. It is important for the designer to be aware of the various design features, which can affect manufacturing costs. A minute change in a filler dimension or shape can cause a large variation in cost due to ease of machining, of assembly, of handling, and of use of standards.

For example: The designing of a rectangular pocket. Corner radii have a major effect on part cost while might not have any bearing on performance. Sharp corners are very difficult to machine and require many operations. It is good practice to use a tool diameter that corresponds to the corner radii. However, if the radii are small, machining the pocket requires a long tool travel path. It might be more economical to use a large tool diameter for material removal and then change cutters to machine the corners. If the radii match the economical tool diameter the machining cost will be minimal.

All this leads to the conclusion that the designer is not an expert process planner and he should leave the filler design to the process planner.

Mr. D: I agree that I am not an expert process planner, and that I, in many cases, do not care at all on the shape and dimensions of the design. But how can I leave part of the drawing by specifying “I do not care”? It is not done.

Mr. CC: There is a method which shows how it can be done, it is called Concurrent Engineering.

The goal of concurrent engineering is to enable an organization to effectively respond to market demands. More specifically, concurrent engineering should facilitate reduced time to market, reduce cost, improve quality, etc.

Concurrent Engineering is directed toward the parallel processing of tasks and provides methods to enable different persons to solve problems under consideration of their specific point of views simultaneously. People of different disciplines must work together in a cooperative manner and understand each other.

In concurrent engineering, all the players from different departments get together to design a product. The design engineers, the production engineers, the quality-assurance experts, the reliability specialists, and the marketing professionals decide together what the product will look like. Fast communication is established between different modules fulfilling different functions in the manufacturing system.

To work effectively in concurrent engineering a core team of 3 to 6 people on the project will work full time and others might work part time. Team members needs training in team building, as well as training in soft skills like communication, conflict resolution and leadership.

The advantages of concurrent engineering are as follows:

1. Reduction in number of design changes which are necessary because of problems of fabrication or maintenance. In the previous structure if no solution could be found to correct the design, it had to be reworked from the beginning.
2. As a consequence of a smooth transition from design to execution to delivery time of a product lead time can be reduced.
3. Reduction in amount of scrap rework.

Mr. C: There is considerable disagreement about the types of organizations best suited for concurrent engineering. Some suggest that a nonhierarchical company with empowered employees provides the most fertile ground. Communication among teams, especially on huge projects, is fundamental to getting the separate projects working concurrently; it's probably better if information doesn't have to travel through too many links in the chain of command. On the other hand, some contend that hierarchies aren't the primary obstacles to concurrent engineering. It's the walls between departments that need to be knocked down, not the organization that needs to be flattened.

Mr. PM: From an engineering perspective, concurrent engineering seems like a logical and simple solution to the problems created by the traditional approach. However, when you consider human effect on team work, it can get messy. If getting people from different departments to work together sounds an awful lot like the cross-functional teams we've been talking about for years, you're not far off.

Mr. M: Concurrent engineering challenges engineers on at least two levels: power-sharing and people skills. The people skills are a delicate issue. It does not mean that engineers are less than socially adept. Most engineers are not good at communication, if they really cared for communication they wouldn't be engineers, they'd be marketing people.

That may be one of the reasons the "throwing it over the wall" approach evolved in the first place. It clearly minimized the amount of time the engineer would have to deal with other people and maximized the time he would spend with the product.

Mr. Q: It seems that the good idea of team work in order to integrate the design and process planning, in other words to DFM - design for manufacturing; DFA - design for assembly, is also a problematic one. The way and means to achieve such an integrated approach in manufacturing can be related to promoting team work among the design, production and inspection departments, which does not necessarily mean working in common groups, but does mean using a common knowledge base to advance simultaneously in different phases of a project.

Mr. CC: It seems to me that several objections were made to implement the concurrent engineering method, mainly as it calls for team work, (or in other words, to design by committee). The main reason was that an experience to work with teams under the method of cross functional leadership was not successful. Similar objections were made when we discussed product specifications. The objections are made not because the idea of concurrent engineering is wrong, but the method of teamwork and working groups. If instead of common human teams, common knowledge bases can be employed, then the same outcome may be achieved without the reluctance.

The computerized committee method as presented in Chapter 7 may be extended to design. The framework of the computerized method, the basic dialog sessions method and the surrounding files can serve design as well. The only modification that has to be made is the content of the files, mainly the technical data file.

The president noted that it seemed that there was no objection to this unknown proposal. May we hear more details about this method?

Mr. CC explained: The technical data file was constructed to assist in specifying product characteristics. The next step is concept design.

The objective of the master product design system in this stage is, as before, to propose solutions, to draw the attention of the designer to the possibilities and characteristics of each solution. In order to supply such data in a practical way, the group technology technique will be employed. The following steps are repeated for each one of the block diagram in the master product design or for a group of blocks that are affected by one another.

The steps of constructing the technical data base are as follows:

1. The first step is to study the topic of each single block of the existing product. A list is made of all designs that were used for that product line, including data on their performance, cost and characteristics. Reference to the detailed design is made including the bill of materials. The different designs are categorized by their method of operation and the parameters that are important to the basic product design. Table 8-1 shows a sample of the technical data file for a weighing mechanism for weighing product.

2. Each type of entry and principle of operation on the list is then analyzed, comparing its performance history to similar design concepts used in other products. The data of the design is analyzed to find out if there were problems with the specific design. New components that have appeared on the market are added to the technical data file so they can be considered in design changes or new designs. The objective of this study is to avoid making the same mistakes that were made in the past, and to find the best design. During this study, proposals are made for upgrading and improving existing products by standardization (i.e. using the same best solution in as many products as possible). Naturally, many design solutions are on the list, in addition to customer satisfactions data, and maintenance department remarks. Such data will be part of the technical data file.

3. Group B file will include the notes from different disciplines regarding each concept of design.

4. A design is selected by comparing the product specifications as defined with the capabilities of each design solution, as appears in the appropriate technical data file. If several designs meet the requirements, secondary design objectives are recorded and analyzed in order to arrive at the best compromise. Instead of making an automatic decision, the system displays a table of option designs with grading according to the system strategy, and the designer makes the final decision. All the necessary data are presented to the designer in order to make an intelligent decision, based on facts and real data.

In this stage each block of the master product design will be treated as a separate design objective. For example: the second block of the weighing product is the load cell. Hence, the present problem on hand is how to design a load cell that meets product specifications.

Table 8-1. Sample of a technical data file for weighing scale

No.	Type	Principle of operation	Remarks	Cost
1	Deflection: Coil spring Leaf Spring Column	mechanical levers	Standard product in different sizes Low accuracy (±60g) low repeatability non linear at the ends	Very Low
2		Potentiometer (linear or rotational)	Medium accuracy (30 g) Medium linearity Easy to connect to display Require power supply Simple electrical circuit	Low
3		LVDT (Linear Variable Differential Transformer	Very accurate (1 g) Require power supply Low power output Needs amplification	High
4		Induction Coil & Capacitor	Very accurate (0.4 g) Require power supply Low power output Needs amplification Complicated electronic circuit	High+
5	Stress & Strain Bar	Strain gauge	Very accurate (0.1 g) Require power supply Low power output Needs amplification Complicated electronic circuit Mechanical simple	Very high
6	Load Cell	Purchased item (usually based on strain gauge)	Very accurate (0.5 g)	Expens-ive

Mr. D was not impressed by the proposed system. I do not understand the uniqueness of the system. That is what we usually do by searching for the most appropriate concept design. We are searching in handbooks, catalogs, we interview salesmen and gathering data. This system is doing the same thing.

Mr. PM agrees that there is nothing revolutionary in the engineering way of operation, concept design is based on option selection and evaluating them. However, I am for adopting the proposed system as it might save a lot of time and speed up design time. With this system the designer does not have to look for data, as he described, by sitting at the workstation the designer will get all the data he needs, while the final decision is up to him.

Mr. PP: I am for adopting the proposed system. Mr. D is right that he gets all the data from books and catalogs. Furthermore, the company's standard books include most of the data required. But I noticed that in many cases the designer just does not bother to go to look for a standard, as it interrupts his line of thought and wastes his time. The proposed system supplies the data without getting up to search for them.

Mr. D was not completely convinced; he argued that the CAD - computer aided design station that he works with allows him to design and draw the product very efficiency, including some design support tools. Why do I need another system?

Mr. PP: There is no argument that CAD is a great improvement and help to the design and designers. However, it does not cover most of the following objectives:

- Search for concepts that meet product functionality
- Assist in translating a concept into engineering design
- Retrieve existing drawings and design by: Concepts, Key, and Attributes
- Automatic change of design
- Automatic design of sub-systems and tooling
- Check and enforce company standards
- Check and recommend design for ease of manufacturing and ease of assembly
- Check and recommend material selection (by inventory)
- Check and recommend tolerances
- Tolerance distribution module
- Tolerance for function module
- Automatic dimensioning
- Analysis of strength & stress and kinematics
- Technology transfer and update (old drawings)
- Automated Process Planning capability
- Automated NC program preparation
- Automated bill of material preparation
- Automated design checking feature (completeness ...)

Mr. CC: It is quite a list; to meet the above CAD specifications the following features have to be available:

- Comparing Work piece drawings in computer memory
- Manipulating Work piece drawing in computer memory

- Updating Work piece drawing
- Have access to auxiliary files such as:
- Feature library
- Machine members library
- Computerized products catalog
- Computerized Material Databank
- Display workplace drawing on monitor
- Fast design presentation in: solid, isometric, picture

Such features can be programmed, but as they are not universal features. I guess that that is why the CAD developing companies do not invest in developing such a program.

The proposed product design workstation is aimed for a specific company with a specific line of products. Its objective is to supplement the CAD system, and not to replace it.

The detail design section can be constructed as follows:

In the previous section the concept design was determined. This decision will guide the designer in the detail design. The concept of performing the detail design is the same as the method used for the concept design and the product specifications. The difference will be in the content of the technical data file. In this stage the data will include references to existing drawings of similar detail designs, and comments on their performance. A comparison of previous designs for similar objectives, results in a master detail design of the product. It will include the assembly drawing (bill of materials), and shape of each item. This serves two purposes. In the initial stage of constructing the data file, the comparison will recognize duplicate designs.

Many designs are duplicated not because the needs of the product, but because they were made by different designers or at a different times. When this happens, existing products can be upgraded by selecting the best design and using it for all other products. Secondly, the best design can be studied in a search for improvements. As time passes, new technologies, materials, and components that may not have been available at the time of the original design can be introduced as modifications if they provide economic benefits, or more important be regarded as a master design for future products. Using the gate method, the group B note file and the check list are valuable tools in selecting the master detail assembly design (bill of material).

This design is regarded as a master design, as it provides a general view of the product and its assembly, but, without exact dimensions and tolerances. Such details might differ for each product objective, and it is up to the designer to make these decisions. Remember that the system is heavily based on a computer program and databases. Because the design is a process of innovation, the decisions are made by the designer. The role of the computer is to draw the user's attention to certain features, to propose designs, and to minimize the need for meetings, but not to make decisions.

The technical data files for this stage will include engineering handbook data such as data on materials and their specifications, local and international standards, tolerance tables, useful equations, and relevant data for screws, bolts, rivets etc. A call for help produces information to assist the designer in selecting components will assist the designer in selecting components. A list of available products is displayed, along with an explanation of the difference among the options and the advantages of and problems with each one option. An additional call for help might provide information on different methods of connection and their effect on cross-section shape.

In addition to the above files and databases, an inventory file and suppliers file are attached. Their purpose is to assist the designer in using standard products and available stock, thereby reducing inventory and delivery delays. Often, there is not one best solution or material, but many solutions. The difference in cost and performance is not always significant, especially if the cost of inventory is taken into account.

The president summed up the discussion by noting that the design workstation sounds a very good idea. However, I am not sure that we have the time and talent to develop such a computer program.

I ask Mr. CC to shop around for such a program.

In the mean time let's adopt the remarks and proposals made regarding design for production, and cost reduction features.

Chapter 9

PROCESS PLANNING

1. INTRODUCTION

The president opened this session reminding that the objective is to work as a team and supplement each other, and not to argue about who contributes more to company success. Our group deals with production technology. The topic of this session is cost reduction by means of process planning. We should concentrate on how to improve process planning and its effect on production costs. Let the process planner describe his job, and then the discussion will be open to all.

Mr. PP: Process planning is the stage where decisions of how a product is to be manufactured are made. The process that transforms raw material into the form specified by the engineering drawing is defined. The process is carried out separately for each part, sub-assembly and assembly of the product. This stage is basically analogous to the engineering design stage, but here the nature of the objective is different.

The prime optimization criterion is to meet the specifications given in the engineering drawings. The secondary criteria are cost and time with respect to the constraints set by company resources, tooling, know-how, quantity required. Some of these constraints are variable or semi-fixed; hence, the optimum solution obtained will be valid only with respect to the considered conditions at the time of making the decisions.

In chapter 8 the interrelationships between design and process planning were discussed. A call to design for ease of manufacturing and ease of assembly including methods of implementations was made.

Mr. PS: I am a “user” of the process planning output, i.e. the routing. I have to schedule tasks exactly as specified by the routing. In many cases the routing, as specified, creates overload, under load, bottlenecks which results in idle resources, increased work in process, delays in delivery dates. I think that this is the main problem with process planning that should be solved.

Mr. PP: It is possible that the outcome of the routine will cause the above effects. However, one should realize that the main objective of my task is to transform raw material into the form specified by engineering drawing in the most economic way. This task is carried out separately for each part and assembly of the product. Process planning is a decision making task with the constraints of company facilities, tooling, know-how, quantity required. Machine loading is not one of my constraints, unless it is specifically defined. Shop floor status is dynamic while routing is static. Routine is set for a life time and cannot be changed whenever overload or underload situations occur.

In many cases, overload occurs because jobs are assigned to the “best” resource available; therefore, this resource might have a long queue in front of it. But this is a result of my optimization criteria and I have no time to change a process plan whenever a queue is forming.

Mr. PM: I am also a “user” of process planning output, i.e. the routing. I have to prepare a master production schedule which transforms the manufacturing objectives of quantity and delivery dates for the final product, which are assigned by the non-engineering functions of the organization, into an engineering production plan. My optimization criteria are meeting delivery dates, minimum level of work-in-process, and plant load balance. These criteria are subject to the constraint of plant capacity and to the routing set by process planning.

The master production schedule is a long-range plan. Decisions concerning lot size, to make or buy, resource planning, overtime work and shifts, and to confirm or change promised delivery dates are made until the objectives can be met. Making such decisions is a tedious and iterative task, it seems that all parameters are flexible and we may change them, except the routine which is fixed and untouchable without the process planner’s consent. True, in cases that we ask the process planner for modifications, he obliges and assists us. Our job is done by computer program, it deals with hundreds items and jobs. We cannot stop the program whenever too much iteration is required to find a solution and to look for the process planner for assistance. This way our job will never end.

Mr. PP: I thank the production manager for being tolerant with my job. It is true that any part can be processed by many different routings. But my job is to select and define the optimum one. I can define a set of routings for each item, but I am sure that a production scheduling computer program is not capable to handle such flexibility. Therefore, whenever they approach us with a request for modifying the routing we do it, in our spare time which is a precious commodity.

Mr. PM: Not meeting delivery dates is not always the process planner's fault. In many cases the delivery date promised by sales to customer are not realistic. The throughput time is longer than the elapsed time between an order's received date and the delivery date. Thus it is impossible to meet that date, no matter what the routing is.

Mr. S: We have no idea and data about the throughput time. In order to get an order we have to improvise and try to satisfy the customer. We assume that production management will be able to meet the promised delivery date, or let them improvise; they may split the order and supply partial shipment on time. The customer might not be annoyed.

Mr. CC: I do not understand why sales do not have the information of what is the throughput time for each product. I can understand it in case the order is for a product that we have not produced before, but for our standard products the product structure and the routing are known.

Mr. S responded that this known throughput time is a theoretical one. It is based on the assumption that it is the only product on shop floor and that there are no items competing over resources. Knowing this throughput time does not help much. It indicates the minimum time but not the actual time.

Mr. PM agrees with this last statement. With the environment where hundreds of items are on the shop floor in some stage of processing, and each item might have dozens of operations, it is very difficult to predict when processing a product will be finished, we know the planning but not the actual performance.

Mr. CC still cannot understand the situation and proposes that sales should consult with production planning as to what delivery date it may promise to customer.

Mr. PM: They may consult, but I do not know if I will have the time to plan this date whenever sales approach me. But even if I predict a date I am not sure that we will be able to meet it. Shop floor control is very dynamic, things are changing all the time, partially due to sales who add orders, change order quantity, insert rush orders etc.

Mr. S did not like the idea of consulting delivery date for each order. I have to sell and not to spend time at the office negotiating terms with production or with shop floor.

Mr. PS: In addition I would like to note the production planning is doing a long range master production schedule, and not short range scheduling. The delivery dates set by this scheduling are for general information only. Sales may use such dates, but should add some safety factors. The master production schedule does not consider any specific processing resource at any specific time. At most it computes the total time required to produce the product mix, on a specific resource group at a specific scheduling period. This information is for long range policy planning and to release jobs for intermittent period planning.

My task is to plan the manufacturing and purchasing activities necessary in order to meet the targets set forth by the master production schedule for intermittent period planning. To plan the number of production batches, their quantity and due date, and to do a detailed resource loading schedule for each resource or group of machines. The decisions in this stage are confined to the demand of the master production schedule, and the optimization criteria are meeting due dates, minimum level of inventory and work-in-process, and department load balance. The parameters are on-hand inventory, in-process orders and on-order quantities.

Both production planning and production scheduling would like to have a more flexible process plan. Working with a fixed routing creates unnecessary difficulties as Mr. PM detailed. A fixed routing is not a must due to technology, as Mr. PP explained. There are many ways to produce a part and product, but the system forces the process planner to select one of them, assume the optimum one, and it becomes THE routing. Why should the process planner decide which routing is the optimum? How does he define optimum?

Master production planning would prefer to have the routing with as few resources as possible. It will make their task easier. Scheduling task would prefer to have as many routings as possible. Let me decide which routing is the optimum at the present scheduling time.

Mr. F: I am also a "user" of process planning output, i.e. the routing. It is my basic tool to predict production cost of any item and product. For me it is just mathematics, I compute the hourly rate of each work station, multiply this rate by the each operation time, and get the processing cost of each operation, summing them up I arrive at processing cost of part and product. Adding material cost and overhead costs, and I think that that is product production cost.

I understand that there is no consideration of order quantity. The scheduling and the planning are fixed no matter what the batch size is. The setup is also a fixed time. Such a procedure distorts the real figure of product processing cost. I recommend that the quantity and the effect of setup will be considered.

Mr. PM: There are two reasons why it cannot be done. Firstly, routing does not supply this data, and secondly, even if production planning would get this information they are not able to use it. The computer scheduling program is not capable of functioning with flexible and alternative routing based on batch quantity. Setup time basically indicates total time without specifying its details. Therefore, there is no way to compare the required set up to that used on the previous operation.

Mr. PP: We devise a process plan for each item separately. We have no information to which product this item belongs, in many cases the same item is used by several products. Moreover, the same item may be required in several sub-assemblies of the same product. Process planning is designed right after product design, well before it goes into production, and well before the company gets orders with known quantity. We have no idea, and it is not our business to know when and in what lot sizes it will go to processing.

Mr. PS: Batch size is determined by production planning, and usually it has nothing to do with order quantity, it is based on optimization algorithm. Their algorithm does not consider scheduling problems, resource waiting for job, or jobs waiting in queue etc. Scheduling prefer to have a flexible routing. Process planning is doing its task when selecting the best resource for the job, hence the books show a minimum processing cost, and long queue in front that resource. This results in idle resources and not meeting delivery date. A routing that is based on several resources, and each with short process time might result in better resource utilization and a better scheduling program, adding alternatives will be a big help.

Mr. F: From your explanation I understand that the cost estimate, by which we base all our financial decisions, is just guess work and probably not even close to actual cost. It is a very dangerous situation, and must be corrected. Is there a way that the actual cost can be closer to the estimated one?

Mr. S: I notice that our costing system is a fixed one, computed as described. There is no effect on selling price and costing of order size and delivery date. It will be a great assistance if sales will have some freedom in offering selling price and realistic delivery date. I suggest that the production scheduling program will be open to all disciplines, there by sales will be able to promise a customer's delivery date without the need to consult production planning personnel. Moreover such a program will assist in responding accurately to customer inquiries regarding their orders.

Mr. FM: This proposal will improve the situation by scheduling with realistic delivery dates, but it still does not improve the delays caused by shop floor disruptions. I propose to let me have freedom regarding the routing, to be able to solve processing delayed by modifying the routing, directing the load from an overloaded resource to underloaded one, whenever possible and economic.

Mr. F: Your proposal will improve meeting delivery dates but might increase processing cost.

Mr. FM: You might be right, but consider that the waste caused by an idle resource, and increase of work in process might compensate the small increase of cost due to employing a second best resource for the job.

Mr. S joined the discussion and supported the idea of having scheduling accessible to all. This option might open the possibility of relating selling price with delivery date. Shop floor control that is allowed to deviate from the routing by using underloaded resources with lower capabilities will increase processing time but might decrease throughput and probably processing cost. Having access to scheduling, and adding a "what if" function such option might be very beneficial to sales.

The president interrupted and said that we are moving away from this session's topic. We will elaborate on such proposals in our next session. We got some ideas of what other disciplines would expect from process planning. Let's have the process planner's remarks and see if such expectations can be realized.

Mr. PP: Process planning determines how a product is to be manufactured and is therefore a key element in the manufacturing process. It plays a major part in determining the cost of components and affects all factory activities, company competitiveness, production planning, production efficiency and product quality. It is a crucial link between design and manufacturing. Process planning activities are predominantly labor intensive, depending on experience and the skill and intuition of the planner.

Mr. C: It is quite a difficult and important task, therefore I wonder that, in spite of the importance of process planning in the manufacturing cycle, there is no formal methodology which can be used, or can help to train personnel for this job.

Mr. CC: There are attempts to develop computer aided process planning programs (CAPP). The accomplishments can be classified into four categories: variant, semi-generative, generative, and AI (artificial intelligence) based systems. The variant approach marks the beginning of CAPP systems and is basically a computerized data retrieval and editing method. The semi-generative or generative approach is a knowledge-based method that partially or automatically generates a process plan according to a product's features and its manufacturing requirements.

Academia spent a great deal on research; more than 156 CAPP systems have been reported. More than 300 papers have been published in this area during the last three decades. Despite the achievements of those CAPP systems, their ability to recognize design specifications and distinguish whether existing process plans suit a new part is still weak and leaves much to be desired.

Mr. PP: Process planning is regarded as an art and not a science. Research in the field of process planning has indicated that all experts have their own expertise and one expert's experience might be different from that of another. It is rare, therefore, for two planners to produce the same process. Each process will produce the part as specified, although different processes will result in different processing times and costs

Mr. CC: The question is who is an expert? By definition an expert is one having or manifesting the knowledge, skill and experience needed for success in a particular field or endeavor, or one who has acquired special skill in or knowledge and mastery of something.

Experience is obtained by practical work, where processes are defined, follow-up is made and corrective measures are taken during production. Experience is gained from problematic processes and rejected parts and corrections are made to obtain a successful result. Very little experience, or even the wrong kind of experience, can be gained from 'no problem' parts.

Mr. C agreed that process planning mostly depends on the expertise of the planner. Therefore, it is rare for planners to produce the same process. Experience is invaluable if it is interpreted correctly. However, usually it is not. It was found that process planners usually make their decisions based upon a global understanding without breaking it down into the individual parameters. They know the problem, they find a working solution, but they cannot pinpoint the controlling parameter that caused the problem. Therefore, they apply the same solution to many similar problems, even if the controlling parameters are not the same. Therefore, this often precludes a thorough analysis and optimization of the process plan and nearly always results in higher than necessary production costs, delays, errors and non-standardization of processes.

Mr. PP: My method of process planning is: based on my knowledge (expertise). My first attempt is to look for a part in our files which is similar to the part in question. If there is one I copy the routine used to process that part to the part in question. In case the part is similar and not identical I modify the routing to accommodate the minor differences between these two parts.

In case that no similar part exists, a list of part features is made. A search for feature processed in the past is made and the process plan used for such feature is adapted for the features of the part in question. Adjusting process operation of the exact shape and size, and the available present resources is made. A list of all operations is made and it is the routing.

Assuming that the previous routing worked well, there is no reason why it would not work well again.

Mr. F: I understand that the specified routing will process the part as specified. However, I do not know if the routing is based on the minimum cost or the maximum production optimum criterion?

Mr. PP: It is the same optimum criterion as used before; I have no idea which optimum it is. Are you sure that there is a difference between the two?

Mr. F: I am absolutely sure that there is a difference; for one extreme example: assembly with industrial robot will be much faster than manual assembly, but the cost of robotic assembly is much higher than manual assembly. Both assembly methods will do the required job, which assembly method is selected?

Mr. PP: I see your point, there is a difference, actually several other parameters affect the decision of selecting appropriate a routing. For example: in case of a rush order and busy period I guess the maximum production criterion will be selected, but for out of season with low orders and jobs, I guess the minimum cost criterion will be selected.

Mr. F: Your logic is reasonable, but tell me how many times you keep planning a routing for the same part? When you plan a routing do you know if it is for a specific order or for general use? Do you know when the routing will be used?

Mr. PP: I am getting involved with things that are out of my field of expertise, someone must tell me which criteria of optimization to use in my planning. Furthermore the batch quantity may also affect my planning, but it is not my decisions to make. I am doing what I was asked to do. Usually I define only one routing for a part and it is used for years.

Mr. CC: What you are actually saying is that the routing used for scheduling is probably not the optimum one, which means that our performance, cost estimating can be improved without changing any of our procedures. All that is needed is to use an optimum routing and our delivery dates, resource utilization, will be improved. Why are we so stupid?

Mr. PP: We are not stupid but practical; we sacrifice efficiency to simplicity and ease of planning. I have no problem to plan a routing for each individual order to fit all parameters if someone will specify the parameters.

Mr. PM: It is not practical to plan a routine for each individual order as no one can specify what kind of routine is needed. Scheduling deals with each order and item of the product mix and they all interfere with one another. A compromise is needed to arrive at a good plan and a compromise can be reached only if there are alternative routings, and it is impossible to tell which alternative will be best suited. It is a loop problem.

Mr. PP encouraged by production planning, said process planning is just a paper planning, by itself it does nothing, but it serves all company activity. I can generate many routings, but it is not for me to tell which one is the best for a specific situation, that is not my job.

For demonstration let examine routings for a part that it's processing requires 8 operations and there are 6 candidate resources (see table 9-1).

Table 9-1. List of operation

The following table list the processing time of each operation

Op	R1	R2	R3	R4	R5	R6	Pr
1	**.40**	.45	1.35	99	1.69	1.24	0
2	**.32**	.37	.88	99	1.22	.54	1
3	**1.24**	1.26	1.77	99	99	1.36	0
4	.77	**.75**	1.26	99	99	.85	3
5	**.29**	.34	1.14	99	2.58	4.7	2
6	**.69**	.94	1.6	2.34	2.57	.92	4
7	**.62**	.67	1.18	99	2.33	1.62	3
8	1.46	1.51	2.02	99	2.36	**1.09**	6
Σ	5.79	6.29	11.19	-	-	12.32	

The following tables list the processing cost of each operation

cost	4	3	1.4	1	1	2	
Op	R1	R2	R3	R4	R5	R6	Pr
1	1.6	**1.35**	1.89	99	1.69	2.48	0
2	1.28	**1.11**	1.23	99	1.22	1.08	1
3	4.96	3.78	**2.48**	99	99	2.72	0
4	3.08	2.25	1.76	99	99	**1.70**	3
5	1.16	**1.02**	1.60	99	2.58	9.40	2
6	2.76	2.83	2.23	2.34	2.57	**1.84**	4
7	2.48	2.01	**1.65**	99	2.33	3.24	3
8	5.84	4.53	2.83	99	2.36	**2.18**	6
Σ	23.16	18.88	**15.67**	-	-	24.64	

The process planner may list the operations and computes the time or cost for processing each operation on each resource. The resource with the minimum value for each operation is selected to perform that operation.

Thus the recommended routing for maximum production will be:

Operation.	Resource	Process operations	Time	Cost
1	1	1 - 2 - 3	1.96	7.84
2	2	4	0.75	2.25
3	1	5 - 6 - 7	1.60	6.40
4	6	8	1.09	2.18
		Total	**5.40**	**18.67**

And the recommended routing for minimum cost will be:

Operation.	Resource	process operations	Time	Cost
1	2	1 - 2	0.82	2.46
2	3	3	1.77	2.48
3	6	4	0.85	1.70
4	2	5	0.34	1.02
5	6	6	0.92	1.84
6	3	7	1.18	1.65
7	6	8	1.09	2.18
		Total	**6.97**	**13.33**

Mr. PS: It is quite a good demonstration. It shows that there is a different routing for minimum cost and maximum production. Why not generate the two of them, and let the user select according to shop load which one to use.

Mr. PM: This method did not consider the transfer of jobs from one resource to another. Such transfer calls for auxiliary preparations such as setup. The recommended routing assumes a very high batch size thus, such extra time and cost is negligible and each operation is processed on the optimum resource. For low quantity such time and cost cannot be ignored.

Mr. CC: Examining the tables in 9-1 it struck me that there is a mathematical minimum and a practical minimum; what I mean is that processing operation 4 on resource 1 (0.77) instead of resource 2 (0.75) will increase the processing time by 0.02 minutes but eliminate the need to transfer the part from resource 1 to 2 and back. Similarly in minimum cost routine the difference of operation 4 between resource 6 (1.70) and resource 3 (1.76) may allow to save transfer expenses.

Mr. PP: When a resource change is needed extra time has to be added to the routine to cover for auxiliary preparations jobs such as setup, let's call it penalty. The penalty value is equal to the penalty divided by the batch size. In case that the batch size is large the penalty maybe negligible, but at lower batch sizes it might be significant. Therefore we should look for part optimization and not for single operation optimization, i.e. select a resource which is not the minimum but by using it no penalty should be paid. For example: the penalty for this part in the minimum cost is 0.7 units. Therefore it will be more economic to process operation 8 on resource 3 instead of Resource 6 (2.83-2.18 = 0.65); operation 6 on resource 3 instead of resource 6 (2.23-1.84 = 0.39); Operation 5 on resource 3 instead of resource 2 (1.60-1.02 = 0.58); Operation 4 on resource 3 instead of resource 6 (1.76-1.70 = 0.06). Thus the routing will be: use resource 2 to process operations 1,

2 and then move to resource 3 to process operations 3, 4, 5, 6, 7, 8. By this method the processing cost was increased by 1.68 units by saving of (6-2) 4 penalties which sum up to 4*0.7 = 2.8.

Thus the recommended routing for maximum production will be:

Operation.	Resource	process operations	Time	Cost
1	1	1 - to 8	**5.79**	**23.16**

And the recommended routing for minimum cost will be:

Operation.	Resource	process operations	Time	Cost
1	3	1 - to 8	**11.19**	**15.67**

Mr. CC: I am impressed by this process planning technique and the results. Actually a simple formula can indicate if it will be profitable to deviating from the operation optimization and move to part optimization.

Mr. PS: This technique is quite new to scheduling techniques and I have to get used to it. Usually in scheduling the setup time is regarded separately from the processing time. The routing specifies setup time and processing time for each operation. Before each operation we allocate (on the resource) time for setup, which is followed by processing time. What you propose means combining these two into one figure. The separation of setup and processing time enables me to start setup independent of the processing, in case that the next operation is on an idle resource and the previous operation is not done yet and thus reduce overall time.

Mr. PP: This technique is mainly a process planning technique, which enables me to consider batch size and setup in making the decision of optimum planning. I seldom use this technique because it takes planning time which I do not have. I demonstrated this technique just to show what can be done. Scheduling may keep on working as usual.

Mr. PM: There is a difference in the term operation between process planning and production planning and scheduling. For me operation is the time that the part occupies the resource. Actually I do not care what is being processed while the part is on the resource. For process planning operation is the individual movement and processing details of what is being done during that time. To make my work easier I prefer to have as few operations as possible. If at the same time an efficient process results, I am in favor of this technique.

Mr. PP: Another technique that we use is by adjusting the sequence of operations. The operation list, as shown in table 9-1, was arranged in an arbitrary manner. The part is composed of features, and each feature was handled separately and its operations are listed in the table. Thus the sequence in the list does not have to be the sequence of processing. However there are some operations that must precede others. By changing the sequence of operations, whenever it is allowed, savings in resource changes can be made and thus savings in paying penalties.

For example: Table 9-1 lists the processing time and cost of each operation, in addition the right column marked as "Pr" indicates priority sequence, which means the operation that must precede the present operation. Priority "0" means that this operation may be processed independent of any other operation. The priority code of operation 5 is "2", which means that operation 5 can be processed only if operation 2 was done. It must not be right after, but at any sequence after operation 2. i.e. the possible sequences may be: R2 - Op. 1 , 2, 5 ; R3- Op. 3, 7 ; R6- Op. 4, 6, 8. Or: R3 - Op. 3, 7; R2 - Op. 1 , 2, 5 ; R6- Op. 4, 6, 8. Or: R3- Op. 3, 7; R2 - Op. 1 , 2, 5; R6- Op. 4, 6, 8. In any one of the sequences only three penalties have to be paid, instead of 7 as shown before, without considering change of sequence.

By the previous method the processing cost was increased by 1.68 units but saving of (7-3) 4 penalties which sum up to 4*0.7 = 2.8. By this technique the processing direct cost is unchanged but savings of 2.8 units are saved by changing the sequence of operations.

Mr. PS: It is all very nice and convincing, but in order to benefit a table of all possible operation processing must be available. As far as I know, usually the process planner evaluates different processes in his mind, and ends up with a single routing. That means that it is a good theoretical techniques but not practical. Process planning is a predominantly labor intensive activity and it is characterized as a bottleneck of production today. They are always very busy and we are satisfied if we get a routing on time, you cannot ask them to spend too much time on any single part.

Mr. PP agrees with this statement. I explained the way that process planning should be done. The table format represents almost an infinite number of possible processes. For a table of N = 8 operations and M = 6 resources, the number of process combinations is $N!M^N = 40{,}320*1.68*10^6 = 1.68*10^{10}$. It will take tremendous amount of time to select the most economical routing but probably it will not be economical to spend such amount of time. What I mean is: suppose that by using my expertise I can process plan a part in 30 minutes and arrive at a processing time of 12 minutes per item. But if I would devote 3 hours to it I may plan a more efficient routing, let's say to process the part in only 10 minutes. In case that the required quantity is 50 units, the total time (planning and execution) will be in the first case: 30 + 50*12 = 630 minutes. In the second case: 3*60 + 50*10 = 180+500 = 680 minutes. This means that it is not economical to spend a long time in search for the most economic routing; it is a function of quantity.

Mr. CC: Why not use a computer program, CAPP - Computer aided process planning, by which the time to generate a routing will be very short, and thus a better process will be planned.

Mr. PP: I wish I could do it, but unfortunately, no such program is available in working order. A lot of papers have been published, proposals are introduced, tremendous effort has been made in developing CAPP systems, yet, it still remains in the conceptual stage, and the benefit of CAPP in the real industrial environment is still to be seen.

Mr. FM: This statement is not quite true, we have some CNC machines which possess process planning capabilities. By specifying a feature the machine will specify the operations to process it; furthermore, it will generate the NC code to produce it.

Mr. D: We are using CAD - Computer Aided Design systems to prepare drawings and there is a capability to process planning some of the features of the design, and link to CNC controller for processing.

Mr. PP: There is such features as you mentioned and they are of great assistance, however, they do not solve the problem of process planning. The CNC can process plan, however, one has to select that specific machine for processing, but it might be that another machine will do a better job.

To generate a good process plan the system must have complete data on the product specifications. The machine is not always aware of them and there is no program to consider them. Therefore the process plan is good for a statistical part and not for a specific one. Unfortunately, present CAD systems were developed to comply with the drafting and minimum design needs. The needs of other stages and disciplines were not specified as an objective, and therefore, CAD systems do not supply neither the data, nor the needed speed of response. CAD systems of today are excellent for displaying part geometry on the monitor, but they are almost useless for process planning.

Mr. CC: The difficulties to construct a CAPP program have been analyzed and methods to overcome these difficulties have been proposed. One of the difficulties is that there is no formal methodology which can be used. It is a huge task to develop such a methodology and it takes a great deal of effort and years of research.

Mr. PM: The technology, as shown, by using a table, is actually divided into two stages: constructing the table, and selecting a routing by using the table. The first stage requires an in-depth knowledge in manufacturing processes. In order to construct a CAPP system that will be available for practical use in a short period of time, it is proposed to divide the first stage into two phases: The human expert process planner defines the process as done today. This process is recorded in the table as the first column. Then the heading i.e. list of available resources is entered. The content of the table is entered by transforming the operation time, as set by the human process planner on his selected resource, to the time that it will take on another listed resource.

This transformation handles technological data. However, as the equations for transformation are straightforward, a computer program can easily be developed to perform this task. The adjustment considers the following factors: resource physical size, accuracy, special features, available power & torque, available speeds and feeds, number of tools, type of controls, handling time etc.

Mr. C: We all agreed that forming a process plan by using a table of possible operations as proposed will result in the most efficient routing. The obstacle of constructing the table was removed by Mr. PM's proposal. In addition to transformation a check for consistency of the of process planner process plan can be made, which might improve the routing. I am for adopting this process planning method.

The president summed up the discussion: Process planning is a passive task, although it is the driving force behind many active tasks. It is advantageous to have an efficient routing, but we have to consider the cost of arriving at such routing. The example made of adding the thinking time of the process planner to the actual processing is quite convincing. A computer program can reduce the thinking time, but as far as I see there is no such program available. We should keep track of developments in this field, and when it will be profitable to use it.

The process planning thinking time and the initial generated routing are highly dependent on the process planner's skill and expertise. In the case of a skilled process planner the extra thinking time for improvement might generate a very low routing improvement.

Therefore I suggest that we should keep on, for the time being, with human process planners, but invest in sending them to seminars, process planning courses and conferences, where by meeting and discussing with their fellow planners their expertise level will increase, and good routings will be generated in a very short time.

Chapter 10

PRODUCTION PLANNING

1. INTRODUCTION

The president opened this session reminding those present that the objective was to work as a team and compelement each other, and not to argue about who contributes more to company success. Our group deals with production technology. The topic of this session is cost reduction by means of production planning. We should concentrate on how to improve production planning and its effect on production costs. Let the production planning manager describe his job, and the discussion will be open to all.

Mr. PM: The purpose of the production planning stage is to plan the manufacturing and purchasing activities necessary in order to meet the targets set forth by the master production schedule. Quantity and due dates are set for each part of the final product.

The master production schedule transforms the manufacturing objectives of quantity and delivery dates for the final product, which are assigned by forecasting or customer orders into an engineering production plan.

The master production schedule is a long-range plan that aimed at serving management. Decisions concerning make or buy, purchasing of addition of resources, overtime work and shifts, and confirm or change promised delivery dates are made.

The master production schedule sets the goals for the production phases of the manufacturing cycle. It specifies what products are to be produced, the quantities, and the delivery dates.

Production planning activities are dependent on this master production schedule; hence, they can be planned and are predictable.

Production planning phase is a mid-range planning with the objective to release jobs to shop floor for processing. In addition to plan external source activities such as what needs to be purchased, subcontracting operations to other shops, subassemblies, assemblies, and raw materials form.

The optimization criteria are meeting due dates, minimum level of inventory and work-in-process, and department load balance. The input is: master production schedule; product structure (bill of materials); routing of each part; delivery date of each product; plant resources. The parameters are on-hand inventory, in-process orders and on-order quantities. The output is a (short term) list of jobs released to the shop floor for processing.

The method to accomplish this task is divided into two phases: requirement planning and capacity planning.

Requirement planning: At any point in time numerous activities are underway in a working plant. There are open shop orders, open purchase and subcontract orders, and items in storage between operations and activities. All of these activities must be considered when converting the master production schedule into production activities.

A working plant is a dynamic environment, subject to many changes and unplanned interruptions, which may lead to the accumulation of unrequired stock. This stock can often be utilized later in manufacturing. The objective of requirement planning is to plan the activities to be performed in order to meet the goals of the Master production schedule, while accumulated stock is taken into account. The product structure is converted to a working product structure. The output of this phase is the input to the capacity planning phase.

Capacity planning - The goal here is to transform the manufacturing requirements, as set forth in the requirement phase, into a detailed machine loading plan for each machine or group of machines in the plant. It is a scheduling and sequencing task. The decisions in this phase are confined to the demands of the requirement planning phase, and the optimization criteria are capacity balancing, meeting due dates, minimum level of work-in-process and manufacturing lead time. The parameters are plant available capacity, tooling, on hand material and employees.

Mr. F: I am using product structure in my evaluation and cost estimation, however, I do not know what you mean by working product structure, please clarify.

2. REQUIREMENT PLANNING

Mr. PM: The working product structure includes only those items that are needed for processing or purchasing. For example: The concept of requirement planning was described in chapter 3 section 3-1. Fig. 3-2 presents 3 levels of a bill of material. Where A is the product, B is a sub-assembly composed of items D and E; and item C. Suppose that there are in stock the full quantity of sub assembly B; then the working bill of material for order A will become A - and C.

Mr. D: Actually there are several types of bills of materials (BOM). The design BOM represents the drawings of the assembled product. The assembly BOM represents the process planning recommendation for the assembly. In several cases in order to improve assembly efficiency the process planner creates artificial sub assemblies.

Mr. PS: I do not understand the need for the working BOM, can you explain? You explained very clearly how to construct it; what is needed (gross requirement) is multiplying order size by quantity per unit as defined by BOM minus what is available. This is not a complicated and a time consuming computation, then why prepare it in advance and create the working bill of materials?

Mr. PM: explained: You are right, it is a very simple computation, but that is not the problem. The problem is with which order to start the stock allocation. Let me explain; the purpose of requirement planning is to plan manufacturing activities accurately by calculating the net requirement in conjunction with the production scheduling. However, if we treat each order of the master production schedule independently, the results will be unreasonable. On hand items will be allocated to orders that require them at a later date, while for other orders a processing order will be issue. This situation is demonstrated in Fig. 10-1. Three end products are ordered: product A for period 9, product M for period 10, and product P for period 11. If we calculate the net requirements for each product independently according to ascending order of due date, we will start with product A. This product requires 100 units of item B in period 8.

Suppose that there is a free stock of 100 units of item B on hand. This quantity will be allocated to product A, and no net requirement for item B exists. Next product M will be dealt with. This product requires 60 units of item B in period 7. Since the free stock of this item was utilized, a net requirement for this demand will result. Next product P will be dealt with. This product requires 40 units of item B in period 6 and 40 units in period 9. Since there is no free stock, a net requirement will result.

These calculated results are unreasonable, since they call for planned orders of item B of 40 units in period 5 and 60 units in period 6 while keeping in stock 100 units not required until period 8. This could result in rush orders or in not meeting the due dates for products P and M. One would expect the calculation to allocate the free stock of 100 units as follows: 40 units in period 6 for product P, 60 units in period 7 for product M, and a planned order of 100 units scheduled to start in period 7 for product A.

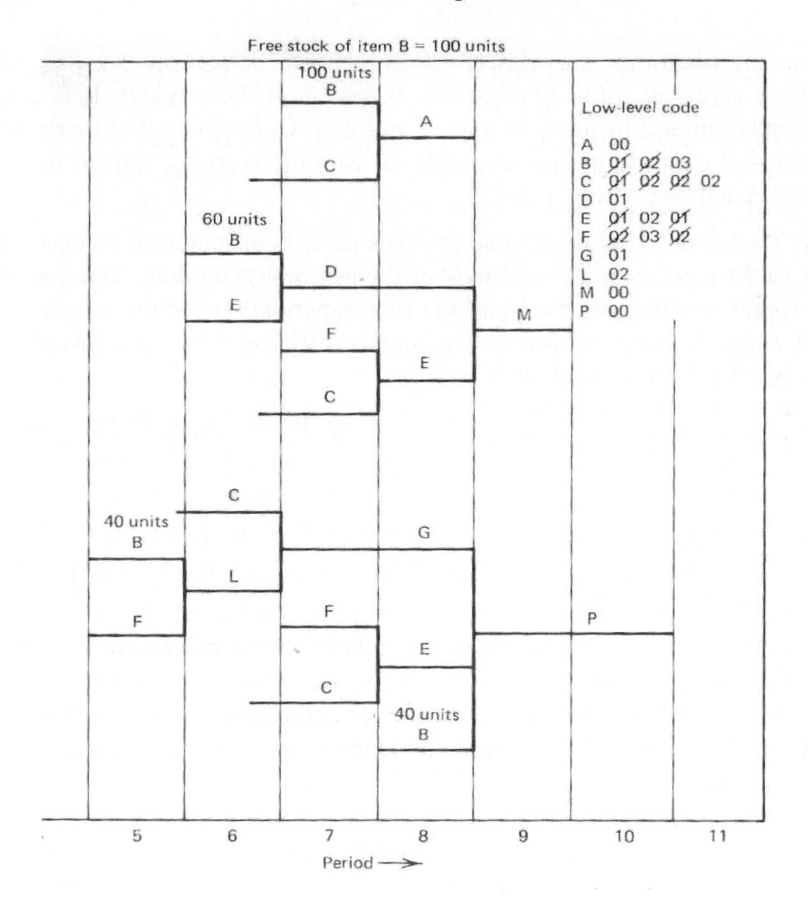

Figure 10-1. Requirement planning for three products

Mr. PS: your example is a convincing one for what not to do. The above example dealt with only three orders that have common items. In practice, the number of such items might be much greater, and I doubt if a human programmer is able to cope and arrive at a decision of the best sequence of orders to treat in stock allocation.

Mr. PM: In order to overcome this problem, the requirement planning calculations are carried out by using a low-level code and not by orders.

Mr. F: asks to clarify what is meant by low-level code.

Mr. PM: explained: The low-level code is an indication of the lowest level at which an item is used in any product defined in the bill of material file. In Fig. 10-1 item B is at level 01 in product A, at level 02 in product M, and at levels 01 and 03 in product P. Thus the low-level code of item B is 03. This code is kept in the item master record.

Requirement planning uses the bill of material file. A table is constructed for each item defined in the bill of material. Initially, all orders of the master production schedule (the independent demand) are read, and the demand is entered as the gross requirement in the table of the appropriate item. These orders are for end products that have a low-level code of 00 or for spare parts, which might have any low-level code.

The free stock (on hand) and schedule of received orders are also recorded in the bill of material file in the table of the appropriate item.

The requirement planning starts by calculating the net requirements of items having a low-level code of 00. The required planned orders are recorded as the gross requirement in the table of the appropriate item; this is done by using the product structure file and the quantity per assembly. When all items on file with a low-level code of 00 have been processed, the calculation will handle items with a low-level code of 01. This process will continue, level by level, until all items in the bill of material file have been processed. By this technique, item B of the previous example will be treated at level 03, at which point the gross requirements of all orders are recorded in its table. Thus the allocation of the on-hand inventory will be logical and according to expectation.

The calculation of the lead time for manufactured items is based on routing data and modifiers, such as transport time, queue wait time, and inspection time.

Mr. PP: The planning sounds reasonable, however, the required period is calculated based on routing data. As discussed before, there are several routings possible, such as maximum production (minimum processing time) or minimum cost. Each routing will result in items needed at a different period, and thus different stock allocation will result. Which routing to use?

Mr. PM: You must remember that requirement planning phase objective is to set due dates for processing each item and its lot size. One does not have to be very precise, it defines a period and not a date and time. I would propose to use the minimum cost routing, it might result in setting due dates too early, but it is better than to have it too late.

Mr. F: Setting a due date too early will probably result in increasing work in process, which defines the objectives of production planning.

Mr. PP: The difference between the two mentioned routings might not be too large; however, multiplying the unit processing time by quantity, the total might be considerable.

Mr. PS: It seems that the proposed requirement planning method does not result in accurate data of stock allocation and setting due dates. In addition I would like to note that the period method, which sounds a good one, does not consider the resource loading. The routing time is for an item. But processing an item is composed of a series of operations; each one might be processed on a different resource. Even if using a perfect routing, there is no guarantee that the time span indicated by routing will be practical. Probably it will be much longer, which might mean a wrong allocation of stock, which will result in increased work-in-process, idle resources, and increased capital tie down in production.

Mr. PM: This phase of requirement planning, i.e. stock allocation does not pretend to do capacity planning, rather set rough priorities of jobs that should be performed in order to meet the master program scheduling set by management. The fine scheduling is done in the next phase, i.e. capacity planning.

Mr. S: Customers usually ask me about the progress of their order, many times they want to change quantities or delivery date, and I have to provide answers. It seems to me that the proposed requirement planning calculates the net requirement and planned orders for each item separately; the product network is lost in the calculations. Therefore it is impossible to examine the effect of request to change the due date of an order. Its logic is concerned with orders as stated in the master production schedule and not with specific customer orders. To keep customers, and good customer relationships we must trace each specific customer order. We must be able to trace the effect of item delays on the delivery date of the finished product. We must have data on which specific customer orders will be affected and what the extent of the delay will be.

Mr. PS: Tracing each customer's order is not only sales interest it is also scheduling interest. Shop dynamics and actual resource loading are out of the scope of requirement planning. In manufacturing, one expects delays in deliveries, excess scrap, unscheduled stock issues and overloaded machines. These factors offset scheduling and could cause a chain effect throughout the product network. The proposed requirement planning is not capable of coping with such a problem which is crucial to sequencing.

Mr. PM: Requirement planning is an excellent tool for calculating the activities required in order to meet the goals of the master production schedule while maintaining a minimum level of inventory. However, it cannot bridge and automatically coordinate the conflicting timing scheduling of the manufactured and purchased items.

It is advisable to use requirement planning only for netting the requirements for each item, while controlling plant daily operations through the capacity planning and job release systems. These systems work with the product network and are able to shift networks backward or forward, to take into consideration the machine load, and to accomplish an accurate, realistic scheduling.

Mr. F: I understand that what you propose is to use requirement planning only for netting the requirements for each item, while controlling plant daily operations through the capacity planning and job release systems. These systems work with the product network and are able to shift networks backward or forward, to take into consideration the machine load, and to accomplish an accurate, realistic scheduling.

That means that stock allocation holds only for the time of allocation. Later in capacity planning and shop floor networks will be shifted, which means that items needed for assembly might be missing in one order, and just waiting to be assembled on another order. Is it not just a waste that increases cost?

Mr. CC: I do not understand the logic of the system. It starts with an order and its bill of materials. Then you break it down to individual items. But it cannot do product scheduling that must run capacity with product structure. Why break the product structure in order to reunite it? And how can you do it? Why not initially do stock allocation on the product structure? You gave an example that when allocating each order independently, unrealistic allocation results, I agree, but is it the only way? Probably not.

Mr. PM: The whole world of manufacturing activities is not an insurance company. Shop floor is a dynamic environment, changes occur all the time. Capacity plans probably hold for at most a day, but that does not mean that it should not be done. True, with all good intentions, there will be waste of material and money, we are trying you keep them down to a minimum, without 100% success. The proposed requirement planning procedure is regarded as the best.

There is a program that can be added to requirement planning in order to link it with capacity planning. It is called "pegging" and it traces individual requirements to their specific source. It provides an upward traceability from component to parent item record, all the way up to the end item requirement stated in the master production schedule.

Mr. PS: It is a necessary feature, as capacity planning must work with networks and be able to move items' due dates in conjunction with other items that are needed for a specific assembly. But as the requirement planning set dates for combined order quantities, it increases the individual order quantity and thus the processing timing. Working with each order separately, the quantity will be the right one, thus the processing time shorter, and thus the order throughput time is shorter.

Mr. CC: FMRP (flexible requirement planning) is a procedure that allocates the items in inventory to the orders and adjusts the early starting dates of each item accordingly while keeping product structure.

It starts exactly as in the method described by converting product structure from level based to time based as shown in Fig. 10-1 (i.e., the length of the connecting lines between product levels will indicate the length of time to produce an item).

The time-based product tree is constructed for all customer orders in a random sequence. For example in Fig. 10-2: product A is an assembly of items 2 and 3, its order specified quantity. Using the process planning table method (as discussed in chapter 9) a routing for assembly A is generated. The assembly time determines the length of this line. Next, items 2 and 3 are treated (in an arbitrary sequence) Let's say sub-assembly 2; its required quantity is computed by the product tree, and it calls process planning to devise a routing for the assembly of items 4 and 5. The assembly time determines the length of line 2.

This procedure continues, covering all orders.

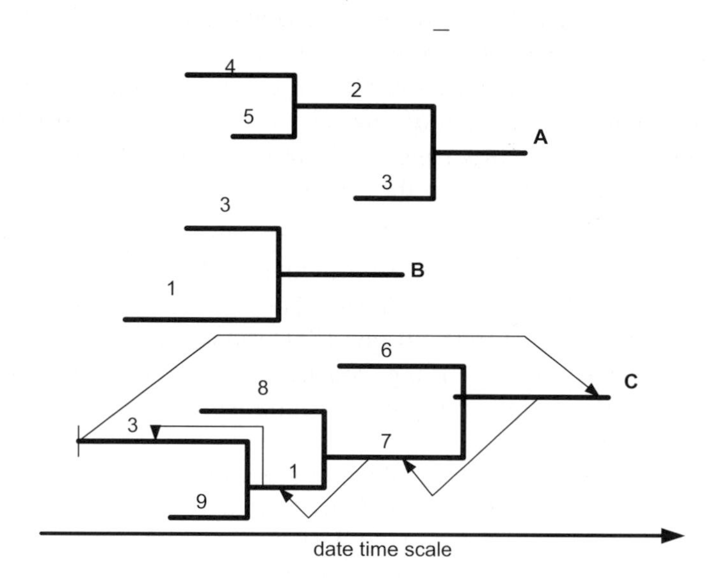

Figure 10-2. Stock allocation sequence of priorities

Now instead of working with low level items as previously advised, the idea is to allocate stock to the critical order. A critical order is defined as the order that's lowest level item has the earliest starting date from all orders, (i.e. the item with the earliest starting date).

When the item with the earliest starting date is determined, the tree for this item is examined in order to find the product and the order; (i.e., the level 0 of that tree) to which it belongs. The inventory is checked to see if this level 0 product is available in stock. If available, it is allocated to this order.

The quantity of this order is reduced by the available quantity, and the product tree is rebuilt with new quantities per item and new early start dates. In this case the early start time of the lowest level item will be changed and another order might become the critical one. The procedure is repeated by scanning all the orders to find the "new" critical item. If the product is not in stock, the availability of the next level item is examined using the same procedure.

For example: Examining fig. 10-2 reveals that item 3 of product C has the earliest starting date; therefore it is regarded as the critical item. However, the allocation priority should be given to the level 0 item of the critical item, since there will be no need for that item if higher level items are available in stock. The chain of the critical path is: items 3 - 1 - 7 - C. Therefore, the system checks if inventory product C is available. Suppose that item C is available in stock for the whole quantity; then all product structure of C is marked as available, and is erased from the time-base product tree. In this example, only products A and B are considered.

In case that only partial quantity of item C is available in stock, it is allocated and the product tree is rebuilt with new quantities per item and process planning is called to devise a routing for new starting dates. Naturally the time base product tree of this item will shrink, and a search for new critical order is made.

In case that no item C is available, a check is made to determine if there is stock of item 7 if no stock is available then item 1 is checked and so on along the product structure.

After each allocation a check is made to find the current critical item. The process continues, till all low level items are marked as treated. This method assures that allocation does not consider the delivery date of an order but instead makes sure that the critical items get a priority.

At the end of the stock allocation step, the product structure includes only those items that have to be produced, or purchased. The working product tree is not similar to the master product tree, as some items might be missing altogether, others might have a different quantity.

Mr. PM: The procedure described above calls for scanning the bill of materials many times, both in up and down directions, (i.e. from level 0 to the low level and vice versa). I assumed that this would take much computer time and therefore would not be practical.

Mr. CC: True, it takes a lot of computations and scanning, but with today's fast computers and modern file management (databank organization) our test shows that it takes just a few seconds to run 23 orders on 15 resources. That means that from this standpoint it is a practical method. While you may consider the low level method that you described to be a fast system, the adding of pegging, which is a must for capacity planning, takes a longer time.

Mr. C: The fault with the low level method is that it breaks the product structure, but it has the advantage of adding items from several orders and thus of increasing item processing quantity, a feature that the proposed method does not have.

Mr. PS: There is a notion that the larger the quantity, the better the productivity. It sounds logical and reasonable. The larger the batches size the lower the setup cost and time. However, I do not know if anyone challenged this notion and proved its validity.

The topic of dispatching rules has been studied by many researchers. As there is no clear definition of good scheduling the following conclusion has been reached:

- Mean flow time is minimized by SPT (shortest processing time) sequencing.
- Mean (and total) inventory is minimized by SPT sequencing.
- Mean (and total) waiting time is minimized by SPT sequencing.
- Mean (and total) lateness is minimized by SPT sequencing.

The mechanism by which SPT reduces the lead time of a job in the system should not be difficult to understand. By giving priority to short tasks, it accelerates the progress of several short jobs at the expense of a few long ones; this was proven mathematically.

SPT is affected by order quantity and routing. Increasing quantity, increase processing time, and thus affects the throughput. The conclusion might be not to increase the quantity by combining items from different orders.

Mr. F: The allocation by both methods depends on the time base product structure, and it depends on the routing used. There are, as we discussed, at least two optimization methods to generate a process plan, i.e. maximum production and minimum cost. Which one is used?

Mr. PM: Requirement planning recommends using the maximum production routing. The reason is that as its name implies, it will result in using the best routing and thus realize maximum throughput time.

Mr. CC: FMRP's objective is to set stock allocation according the criticality of the order. Anyhow, the routing that will be employed in the next stage, i.e., the capacity planning must not be the same as the one used for constructing the time based product tree. It is important to check if there is a slack, or if the low level item falls in the past, i.e., below the current day. Therefore, FMRP might recommend to start with minimum cost, and if delivery falls in the past to switch to maximum production routing.

Mr. F: I understand that the FMRP method is more precise in stock allocation. Keeping the product tree as part of the scheduling and processing can eliminate waste of inventory and cost. It can respond quickly to disruptions, change of orders, and lends itself for customer relations by being able to report at any instant the progress of each order. Requirement planning is just a planning stage, but its planning may cause indirect waste. In case that the system cannot respond fast enough to customer order changes, dead stock might accumulate. In case of wrong planning, items might be missing for assembly or processing and thus idle resource will occur, which will result in not meeting promised delivery dates. Capacity planning and shop floor stages are allowed to shift networks, which means that item needed for assembly might be missing in one order, and just waiting to be assembled on another order. Is it not just a waste that increases cost? I estimate that substantial savings can be realized by using FMRP.

3. CAPACITY PLANNING

Requirement planning specifies the activities to be performed in order to meet the goals of the master production schedule. It plans both purchasing and production activities, taking account of requirements, but disregarding such manufacturing details as machine loading and shop dynamics. It sets objectives that must be transformed into a detailed loading plan for each resource or a group of machines in the plant. As distinct from this, capacity planning is the details planning phase; it is a scheduling and sequencing task. On the basis of the scheduling, it initiates productive activities by the issuing of orders to the shop floor, thus its main objective is to plan order release to shop floor, which is the execution phase.

The major objectives of capacity planning are:

- Meeting delivery dates.
- Keeping to a minimum the capital tied down in production.
- Reducing manufacturing lead time.
- Minimizing idle times (machine out of work) on the available resources.
- Providing management with up-to-date information and solutions.

The capacity planning system encompasses the following:

- Planning the capacity required at each work center and helping to allocate the machines and manpower required to meet the goals of the master production schedule.
- Controlling the level of work-in-process by regulating the rate at which orders are released to the shop floor.
- Helping to reduce manufacturing lead times by reducing the time a job must spend waiting for a machine.

- Planning and minimizing queue lengths to help ensure that machines and personnel will not run out of work.
- Determining how much work can be transferred to alternate work centers in an effort to reduce overloads or fill idle capacity.
- Analyzing remaining overloads and underloads to determine which orders can be subcontracted without causing idle time in other work centers.
- Assisting in making short term capacity adjustments by planning over time, adding temporary extra shifts or releasing work to subcontractors.
- Leveling the planned load on each machine center (in certain instances), thus reducing idle time, overtime, subcontracting, and amount of manpower movement between work centers.
- Determining which orders should be released earlier to prevent idle time.
- Accurately estimating the completion time for every shop order and customer order.
- Planning the sequence of operations to be performed at each work center and providing a work sequence list for the foreman and for other phases.

Mr. PS: explained (by example) the basic scheduling technique used to accomplish the above objectives.

Fig. 10-3 shows the scheduling of three items (jobs). Each operation is marked by three digits: The first designates job number; the second, operation number; and the third, machine number. Initially all three items are backward scheduled. The full lines in part A gives a cross section of order status.

Part B of the figure gives a cross section of the machine load for the backward scheduling of Part A. It shows that on days 73 to 76 machines 2 is overloaded: jobs 1 and 3 require the machine at the same time. A similar situation occurs on that machine on days 89 to 93 and on machine 3 on days 82 to 86 and days 100 to 102. To balance these loads, the slack time may be used. Since job 1 has no slack at all, any change in its latest start will result in a late delivery. Job 2 has a slack of 15 days and job 3 of 10 days.

Part C of the figure shows the machine load cross section after the overload has been resolved by pulling jobs forward. It is based on latest start date loading and considers available capacity.

The dashed lines in Part A show the planned cross section of the jobs. Job 1 does not have a slack and is therefore unchanged. In such cases, early start = latest start = current date and early finish = latest finish = delivery date.

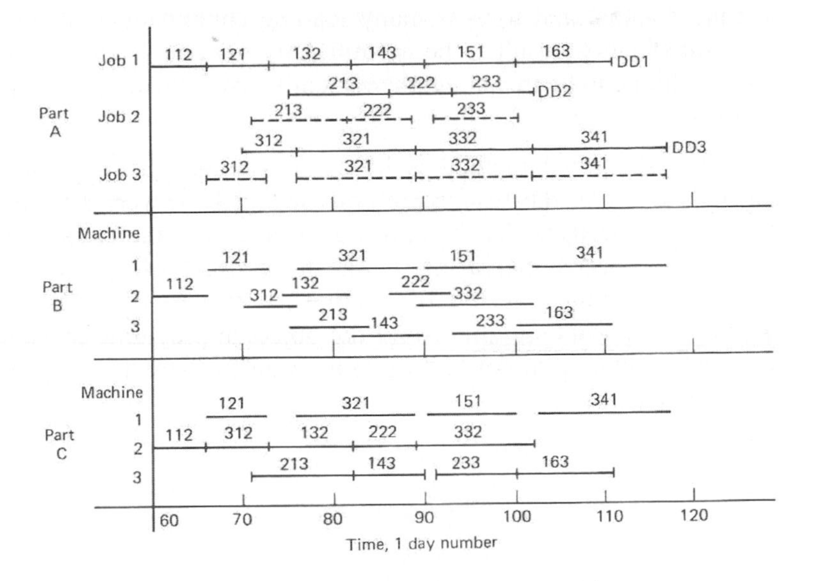

Figure 10-3. Scheduling technique (of three jobs)

Job 2 used 4 days of its slack in order to balance the machine load. It has a new slack of 11 days, the latest start being on day 71. The job is scheduled to be finished on day 100, two days before the delivery date.

In scheduling to finite capacity, all item operations and machines are linked together, and the meaning of the slack is changed: The item slack and operation slack do not coincide. In the example shown in Fig. 10-3, the scheduled latest start of task 213 is day 71. However, machine 3 is unoccupied, and this task may start on current date = early start = day 60. Hence, this operation has early start = 60 and latest start = 71.

The second operation (task 222), due to the fact that machine 2 is occupied, can start only on day 82, which is its early start. It is scheduled to be finished on day 89, two days before the required date. Hence, this operation has early start = 82, latest start = 84, early finish = 89, and latest finish = 91.

The third operation (task 233) has a scheduled latest start of day 91. However, by machine loading one can see that machine 3 is not occupied on day 90, and hence the operation can be pulled forward to start on day 90. The result is early start = 90, latest start = 91, early finish = 99, and latest finish = delivery date = 102.

The slack value may be positive, zero, or negative. A zero slack is sometimes referred to as critical, while a negative slack is called a delay. When working with networks, there is a third type of slack - network slack.

The overall manufacturing elapsed time is referred to as the manufacturing lead time. Scheduling of the items in Fig. 10-3 was done manually. One looks at the diagram and tries as many loading combinations as needed to obtain a satisfactory result. The terminology that has been introduced enables scheduling to be treated mathematically, thus allowing a computer to be employed.

Mr. F: The described method is based on pulling jobs backward, and pushing jobs forward, the results are convincing. However, this algorithm contradicts the objectives that it supposes to meet. Let me explain; pulling jobs backward, means increasing work in process, while pushing jobs forward jeopardizes the objective of meeting the delivery date. Furthermore, such pulling and pushing might disrupt the object of providing management, including sales, with up to date information. Can't you find another method that will meet the goals?

Mr. PS: You are right, some of capacity planning objectives conflict with each other. To minimize the capital tied down in production, the work should start as closely as possible to the delivery date; this will also reduce the manufacturing lead time. However, this approach will increase resources idle time in an environment in which resources are not continuously overloaded. The problem of capacity planning is to arrive at the best compromise between all objectives.

Mr. C: It is all very nice, it solves the problems of meeting due dates set by requirement planning. Moving jobs forward or backward reduces the order network to a useless form. A solution was achieved for one job, but probably created problems with other jobs of the product structure. Attempting to follow the product tree and adjust other jobs of the same order will result in an endless loop of job movements, with no success. It is just a waste of time. One might decide to try only two loops and then to give up.

Mr. PM: The main problem with capacity planning is how do we measure and define good scheduling? What are we actually trying to accomplish? There are many criteria by which one can define the goals of scheduling, such as:

- Minimum level of work-in-process.
- Maximum number of processes completed.
- Maximum number of jobs sent out of the shop.
- Minimum number of processes completed late.
- Minimum average lateness (tardiness) of all jobs in the shop per period.
- Minimum queue wait time of jobs in shop.
- Minimum number of jobs waiting in shop.
- Maximum shop capacity utilization.
- Minimum number of jobs waiting in queue for more than one period.
- Minimum size of jobs waiting in queue for more than one period.

They all are very important but one cannot satisfy them all.

Mr. CC: A great deal of research work has been done in an attempt to optimize production. We find fascinating scheduling and sequencing theories, dispatching rules, and so on. They all use advanced mathematical techniques in order to solve production problems and arrive at optimum operations. Sophisticated computer programs are available with built-in algorithms to decide when to split a job between two or more machines, when to apply overlap of operations, when to reduce transfer time, when to use an alternate machine, how to resolve competition between operations for facilities, and so on. Nevertheless, with all these, the goal is not usually achieved.

Mr. FM: In a plant there are, and will always be, "better machines" that will be overloaded, or the bottleneck in production, while other machines will usually be underloaded. Overload means that there are larger queues waiting for the machines, and these queues increase work-in-process. They are usually resolved by pulling jobs forward, which might result in not meeting due dates. Underload means that in order to balance the load and supply work to operators, the jobs are pulled backward, thus increasing work-in-process.

A bottleneck is usually created by the process planner, who is doing his job honestly, i.e. selecting the best machine for the job. He has nothing to do with scheduling, he does not know when and with which product mix the specific job will be scheduled.

Mr. PP: Any item can be produced by any available resource. It can be produced with 5D DNC machining center, a universal machine, and industrial robots or manually with the aid of a chisel and a file. The cost and lead time required are a function of the process used.

I receive a drawing of an item (not a product) and am asked to devise a process plan for it. Naturally I define the best process possible, using the most appropriate resources. The process is defined independently of the capacity planning, some routings are used years after they were defined.

Most of capacity planning problems and complexity are a result of the solution approach and not inevitable. Most of the disruptions are a result of the rigidity of the system where decisions are being made too early in the manufacturing process. By a different approach, one that will introduce flexibility to the manufacturing process, most of the disruptions will be solved by elimination. The proposed method, (as discussed in chapter 9) recommends that process planning will not make decisions but rather generate alternatives, in the form of table. Capacity planning instead of breaking product structure by moving jobs forward and backward, will choose an alternative resource to perform the bottleneck operation.

Mr. F: I like this idea, let's check it with an example; suppose that the part data in table 9-1 is ordered with a quantity of 50 units. Checking the table the best time to process this item is on resource 1 and it will take (5.79-1.46) * 50 /60 = 3.6 hours. That means that another item that its process plan calls for using resource 1 will have to wait for at least 3.6 hours. Probably it will not be the only one and a long queue will build up. However, instead of standing in line, alternative routing will be chosen, such as using resource 2. The processing time will be (6.29-1.51)*50/60 = 3.98 hours. Which means that it will take (3.98 - 3.6) = 0.38 hours longer to process but it will save 3.6 hours standing in line while resource 2 is idle.

Mr. CC: Most of capacity planning problems, difficulties and disruptions are a result of the solution approach and the rigidity of the system where decisions are being made too early in the manufacturing process and they are not inevitable. By a different approach, one that will introduce flexibility to the manufacturing process most of the problems might be solved by elimination.

Mr. PM: Following your remarks I would like to propose a system that is a direct extension of the FMRP, where a working product structure was defined. Machine loading is forward planning, giving priority to the critical order. The critical order is determined following the same procedure as described in the allocation section. Thus machine loading will start at the lowest independent item in the product structure of the critical order through its sub-assemblies up to the product. However, all the items for a sub-assembly have to be available before the assembly can start.

Machine loading employs a load profile table; were each available resource is represented by a column and each period is represented by a row. The data in each cross section slot of resource and period indicates the state of the resource at that period. If the content of the slot is a data, it indicates that the resource at that period is occupied. The data indicates the processing operation of a specific item. If the slot is blank (empty) it means that the resource at this period is idle.

The loading procedure is as follows:

Start loading according to the priority sequence. Store the item name and quantity to be loaded. Call the process planning table with the item name and quantity and retrieve a process. The retrieved process indicates the number of resources, the name of each one, their sequence and the processing time for the operation (including set-up and penalties). Start with the first operation, multiply its time by the quantity and divide by the period scale. Determine how many periods are needed for this operation. Determine according to the sequence of processing operations, and item dependence (as indicated by the product tree), at which period the processing may start. Turn to the load profile table and run over the row of the appropriate resource. Check if the resource is idle at the required starting period and the number of periods required. In case that the resource is idle at the required time span, the name of the item is inserted in the rows of the load profile table.

In the case that the early available idle period is too far from the early start period, the process planning table will be called to generate an alternative process. The alternative process will attempt to reduce the waiting time by employing a different resource which is idle at the required periods and that the change is economical. The process planning table method generates an alternative process by blocking the occupied resource i.e. ignores this resource in generating a process. If a machine is known to be idle, the "forced process plan" feature of the process planning table method will generate a process by using this machine, the economics of using that process plan are examined. This process may be continued until the available space is found.

Mr. F: It is all very nice, however, it increases the costs, while we are looking for costs reduction. Assuming that the first option is the most efficient process plan, then any alternative increases processing time and cost.

Mr. C: There are misconceptions regarding processing optimization. There are several meanings of optimization: individual operation processing, item processing, product mix processing, (and business optimization). In each case it has another meaning.

Individual operation refers to it as a stand alone optimization. Item optimization refers to all individual operations required to produce an item. It was explained that to transfer operation from one resource to another, a penalty has to be paid. In case that the penalty is larger than the saving in processing the individual operation, then it prefers to process an operation inefficiently in order to get item efficiency.

Product mix optimization refers to all orders that are needed to be produced at a certain time period. In case of jobs competing over resource, it is sometimes preferred to produce an order inefficiently in order to get product mix efficiency.

It's true that the total will be theoretically increased, but it prevent practical expenses of not meeting delivery dates, increased work in process, rejects items etc.

Mr. PR: If I understood the proposed method correctly, it imitates our purchasing methods. Basically, the purchasing can be regarded as a manufacturing department where job orders are issued and items ordered are supplied. The decisions that procurement personnel have to make in purchasing concerns the selection of a supplier, subject to the optimization criteria of quality, quantity, delivery date, and cost. These optimization criteria may conflict with each other (e.g., cost versus quality or delivery date versus quantity), and procurement personnel must find the best compromise, taking into account the constraints of the production plan. Purchasing method is to contact several suppliers with a request for quotation. Decision is made by analyzing the quotations.

The analog to capacity planning is by regarding the different resources as suppliers. A request for quotation to process a single operation or assemble an item is made to all resources; some may respond that they are busy, or unable to perform, while others will specify the processing time/cost that they offer. The capacity planner, (procurement personnel) will make a decision as to which offer to accept.

The process planning table can be regarded as a summarized response of all agents (resources) to the request for quotation.

Mr. PM: The system that I introduced integrates the load profile table and the process planning table, and is a natural extension of the stock allocation. It is one system that works with the product tree and load to finite capacity.

It meets the capacity planning objectives of meeting delivery dates, keeping work-in-process to minimum, and resource utilization to maximum. True processing cost of any individual order might be slightly increased (sacrificed) for the benefit of meeting other objections.

The president is impressed by the proposed method. If I understood correctly, it is a concept that encompasses several disciplines of manufacturing management. It introduces technology to all disciplines of the manufacturing process, leading to increased efficiency.

I asks Mr. CC to look into such a computer program and Mr. F to evaluate the declaration that it's worthwhile to increase processing cost, in order to reduce business cost.

Chapter 11

SHOP FLOOR CONTROL

1. INTRODUCTION

The president opened this session reminding those present that the objective was to work as a team and supplement each other, and not to argue about who contributes more to company success. Our group deals with production technology. The topic of this session is cost reduction by means of shop floor control. We should concentrate on how to improve production control and its effect on production costs. Let the shop floor manager describe his job, and the discussion will be open to all.

Mr. FM: Capacity planning is a simulation of what is likely to happen on the shop floor. It attempts to schedule the jobs with respect to the current production plan and existing manpower and resources. Theoretically, capacity planning has scheduled the work to the last detail, and the shop floor foreman simply has to carry out this plan by assigning jobs to their operators. In practice, it never works this way. In spite of the fact that the load was balanced and all the competition for capacity resolved, the foremen still face the problem of job competing for capacity.

Capacity planning is good for the purpose of releasing orders to the shop; however, it cannot take into account any unplanned interruptions that occur. In the shop, unplanned occurrences take place: machines break, tools break, operators do not show up, actual operation time doesn't work out as planned, the previous operation is not finished on time, a lot is rejected or the foremen, for their own reasons, change the planned sequence. All of these cause changes in the implementation of the capacity planning program. Usually, the life of a schedule is no longer than a day, after that the capacity plan will probably not resemble reality at all.

The capacity planning simulation disregards the unplanned interruptions, that is, an operation is available for scheduling when the latest finish and the interoperation time of the previous operation was due. In practice, an operation can be loaded only when all previous operations have been completed and the components are available in the queue of the machine.

The actual allocation to the individual resource is made on shop floor by the foremen. They know best what is going on in their departments, the particular skill of each operator, the tolerances on the machines, and so on.

When a resource becomes free, a decision must be made about its next operation. This decision has to be made on the shop floor within a very short period of time; otherwise, resources and operators are idle.

Many methods have been proposed to guide and assist the foreman in making job allocation decisions. Paradoxically, this problem defies algorithmic solution for only a few simple situations. Other methods are based on operation priority or dispatching rules and others are based on bottleneck optimization methods, etc. and usually on the skill and expertise of the foreman.

Mr. F: Considering the shop floor foreman's job description, it seems to me that the capacity phase is superfluous and redundant. Probably it is possible to proceed from the master production schedule, straight to job release. Let shop floor work by objectives and not by detailed schedule. Can the production manager give us a brief description of how other companies control shop floor scheduling?

Mr. PM: There is no unique method, one may find all sorts of ideas, and methods, and I guess that it depends on the company products and personnel.

Many companies 1eave the daily scheduling to the foremen. On the other hand, some companies do daily scheduling by computer and force the foremen to follow that plan precisely, while others regards the plan merely as a guide to the foremen, while others might establish dispatching rules to guide the foreman.

Mr. FM: As odd as it might sound, I am for doing capacity planning. It does not mean that I have to follow its planning to the word; I must have freedom to manage my department. But freedom does not mean chaos; a product is being processed by several departments, and we must coordinate production. I have to know the linkage of these jobs to with other departments. This coordination is done by capacity planning.

Mr. PS: I agree with Mr. FM that he must have freedom in managing his department, but it should be limited freedom. Job released supply to the foreman a list of jobs to be processed in a certain period. The list might cover a period of a week or two. This procedure gives the foreman an idea of what jobs will be required in the coming days. But it does not mean that he may choose any job to process at any time within this period; otherwise the delivery date will never be met. The freedom, that I propose, is that if for one reason or another, the next job on the list cannot be processed, it should be skipped to the one next in the list.

Mr. FM: I agree that the jobs should be allocated according the sequence indicated on the released jobs list. Unfortunately, it is not always possible. Beside the "normal" disruptions there are disruptions caused by dispatchers asking for rush jobs. Urgent jobs are a pain in the neck; running jobs must be interrupted in order to clear the machine, a new setup must be done, the batch size is broken. It causes waste of processing time and cost, and above all, it is impossible to meet due dates and it annihilates the capacity plan.

Mr. PM: Rush orders are mainly because sales promised customers that we can supply the orders without checking with us if the promised delivery date is a realistic one, and if it will cause delays with other orders. Several times the time span between the order date and the promised delivery date is shorter than the throughput time of the order. What we are trying to do then is to aim for a minimum delay on this order, and possibly on other orders as well.

Mr. S: When a customer is willing to place an order with us I cannot refuse it. It might be that I get carried away and promise the delivery date that he asks for. I count on you to be able to oblige. It is not good business to reject customers.

Mr. PM: It is also not good business to make promises that you cannot deliver.

Mr. F: I understand that rush jobs create increase processing cost. It causes extra setups, increased work-in-process, it may cause working overtime or extra shifts, extra changes in routing. Who is going to pay for that extra cost? Someone should. I would propose to shift that cost to the customer, by increasing the normal cost. Maybe we should have a cost as a function of delivery date.

Mr. S: It is all very rational but I am not sure that it can be done. We have published our price list; it does not mention that it is for a specific time delay between the order date and the delivery date; I think that it will cause customers to regard this as a kickback and unfair business and we might lose customers. Besides that how can we evaluate the extra costs?

Mr. PS: It is a good idea, but it is difficult to assess such extra costs, it is a function of workshop load which is a variable, some urgent jobs might save costs by using idle resources. We should include all shop floor disruptions as overhead, unless a method of computing extra cost, or individual batch cost as a function of delivery date can be formulized.

Mr. PM: Rush jobs and disruptions are not always the fault of sales. They may be caused by problems in one department that reflects on others. Processing a product is a chain of activities in several departments; production scheduling must coordinate the flow of material, and the Production Planner does it to the best of his knowledge. However, in case there are problems (or disruptions) in one department, the synchronization must be updated. Thus rush orders or changes in items' due date might be made.

Mr. FM: I may recount some other reasons for changes in the released jobs. I do not say that it can be avoided altogether, there are, as pointed out, unavoidable occurrences. However, there are some system faults that may be avoided. In several cases the capacity plan works with inaccurate data. It may be that the product structure is not up to date, using design product structure and not production product structure. Similarly, the routing files are not up to date as to what really happened on shop floor, or planning with infinite capacity, or the actual resource is not the one planned, due to resource being idle due to breakage, maintenance, or waiting for setup approval by quality department.

There should be better communication and interaction between shop floor and capacity planning, or even better joint effort in scheduling and planning.

Mr. D: We design the products and the items and we believe that we are doing a good job. The outcome of our design is a product structure and detail drawings. They should be the basis for production management. In case the drawing and the product structure that we defined are not realistic, the question is: who changed them and on what authority? We are open-minded and if someone thinks that changes are in place, we would like to hear about it, discuss it and make a formal change. It is impossible for production planning to change our design without updating the official drawings.

Mr. PM: Formal changes are the correct method to keep product data management (PDM) files. Only one authority is allowed to update changes. However, when a rush order arrives that requires some changes, we first produce to the customer's specifications, and then (if we do not forget) call PDM for update.

Mr. C: You both are acting in favor of the company interest; however, production management is solving a specific problem, while creating long term problems which obliterate the effort and objectives of their own interest of planning a realistic scheduling. There must be procedures that satisfy all criteria. If the changes are one time for one order, then it is one thing, that should be allowed, but it should still be report to design. However, if is a permanent change, it should not be allowed, and the change should be made through the proper channels.

Mr. FM: There are two types of changes: design changes and processing changes. I never do design changes on my own; I always follow production scheduling release jobs, or dispatcher instructions.

Following the formal production and processing plan without changes will cause profit losses to the company. Due to the shop floor dynamics I must find and perform alternate solutions to a specific situation, otherwise, I have to stop production, leave resources idle, and ask for a formal solution from design or production scheduling. Shop floor personnel know best what is going on in their departments, the particular skill of each operator, the capability and tolerances of each resource, and so on. I believe that we are better qualified to solve a specific situation, which is not a permanent change, but rather a first aid change. We must have freedom to alter scheduling or routings for to the benefit of the company.

Mr. PS: I wish I could also have freedom to deviate from the formal data that I have to work with. The formal data may keep clear the order of procedures, but it also robs the company from possible efficiency. In many cases I wish I could get a better loading by minor changes in routing, but it is not allowed by present day technology. For example: to solve overload or bottleneck machine by moving jobs to another resource that is not specified in the routing. I know that I can reduce lead time and processing cost, but I am not allowed to do so. You might say: who says that you cannot do so, just ask the process planner to change the routing, he probably will do it. Scheduling several hundreds jobs is an iterative task, a change in one job might cause difficulties in another job, I need freedom on the spot, and not bothering process planning several times a day asking for modifications.

A simple example of what I am talking is given in Fig. 11-1. Amounts of idle time will occur at various points within the schedule. Minor changes in working condition as specified by the process planner may by used to eliminate such idle time.

Part A operation A01 calls for machine R1, while operation A02 calls for machine R5, and operation A03 calls for machine R1. Part L operation L01 calls for machine R5, and operation L02 on machine R1. With an unaltered routine R1 is idle from the time A01 is done till the time that L01 is done. There is not much sense in finishing A01 ahead of the needed time and then wait on shop floor for the next operation. To eliminate the waiting time of item A, cutting speed may be reduced (work inefficient) to such an extent that there will be no idle time on machine R1. Technologically it is possible; the saving will be in tooling cost. Another case is that machine R1 is idle from the time that operation L02 is done till the time that operation A02 is done. It might be possible to speed up the performance of operation A02 in such a way that it will be done just as L02 is done. In this case (if it is possible) the tooling cost will be increased while processing lead time will be reduced. The foreman may also initiate such changes when he sees fit.

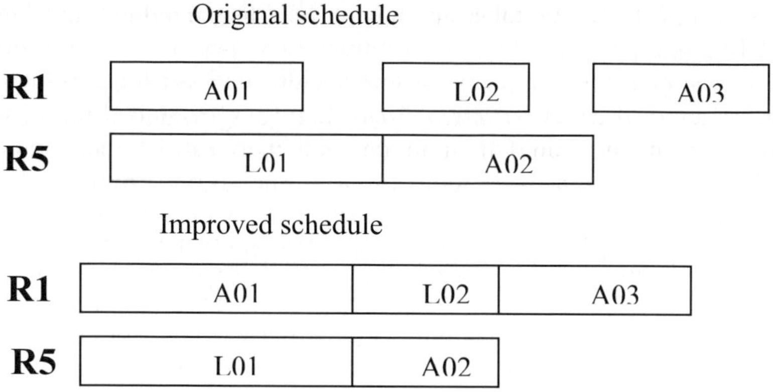

Original schedule

R1 A01 L02 A03

R5 L01 A02

Improved schedule

R1 A01 L02 A03

R5 L01 A02

Figure 11-1. Integrating technology in scheduling

Mr. PP: It seems that you are taking my routing very seriously. Of course if someone will ask for changes, and it will make sense I will be willing to change. Processing an item may be done by many routings, fortunately, or unfortunately, I have to select one and call it our company routing. About the processing time, actually there is no such thing, same as in design, each data has a tolerance, in design there is a method of how to specify it, but there is no way to specify it on routing. A manual operation has a span of 75% to 125%, while I have to specify the mean value of 100%. A human being does not work at a constant pace; it depends on his mood, the time of the day, the shift, the problems at home and many other factors. The processing time on metal cutting machines is a function of tool life and it also has a very wide range. The time is predictable, but may be control by the operator. The speed of movement of industrial robots arms is also programmable and can be regarded as a variable.

Therefore jobs may be finished head or behind schedule, which might leave idle time or waiting time on the next work station. I am for allowing production schedule and the foreman to alter routings without asking permission each time, assuming that they are responsible and skillful in their tasks.

Mr. PS: This is all very well, but all scheduling programs and methods work with a specified operation time. Even if process planning will specify a time range, I have no tool to use it, unless I finish the schedule and search for idle time between jobs, and then I might use the time range to eliminate the idle time. I like to have job time range and a list of alternatives, but I must have tools to take advantage of it.

Mr. FM: I do not have the problems of production scheduler as I work with a limited number of jobs and a limited number of operators at any one period. I am aware of any problems in my shop; actually it is my job to find a solution right on the spot when they occur. Knowing the time range of any job may assist me to take advantage of such freedom and overcome disruptions.

I must know what is going on in the company and other departments. I have to know where my finished items are going to and from where I am getting new items. Can I trust to get the items on time and what is the importance of delivering my items on time? I need information.

Mr. PM: Information is vital for daily sequencing and for capacity planning. It is the basis for many other applications, such as costing, salaries, incentive pay and absentee control. The required frequency of receipt of this feedback is a function of the application. For daily scheduling it must be processed daily or hourly, while for salary and costing a week or a month will probably suffice. However, for reliability purposes it is recommended that it be processed daily or even in real time by data collection equipment. Data collection terminals and equipment replaces manual timecard entry, it enables to integrate system modules in real time to update job tracking information.

Mr. FM: Data collection terminals may assist in getting information on the progress of production in any department. May I add that the production schedule plan may supply data on what is planned to be done. Comparison of the two may serve as exception report that will guide the foreman, while giving him complete freedom to make changes that will remedy the situation and bring production plan to as close as possible to the original plan.

Mr. F: Following the discussion which aimed at giving the foreman freedom to alter and change scheduling and routing, I wish to repeat my previous suggestion to let shop floor work by objectives and not by a detailed schedule. If there is no detailed plan, there are no changes of plans and there will be no spending time and cost to find the exceptions. Why not go back to the definition of the production schedule objective, which was to release jobs to shop floor.

Mr. PS: Job release must be practical, otherwise all jobs might be critical, and no order will be finished on time. To arrive at a practical plan we must plan with all the details. May I suggest that, based on the detail planning, jobs be released to shop floor, and will include a list of jobs that has to be finished at a certain period. However, the foreman must not follow the planning details that were the basis for the release job list, he is free to manage his department as he sees fit. If he argues that the list is not practical, then production scheduling can show him and prove to him that it can be done, if he follows the scheduling detail plan.

Mr. F: I still do not get it; I assume the production scheduling is doing a very good and efficient job in planning orders, and that its plan is the optimum. Now we allow foremen not follow this plan. That means that we are not being efficient and that processing costs will be increased?

Mr. PS: There is no contradiction and decrease in efficiency. The job release stage was done in the office with stable conditions. Conditions on the shop floor are dynamic, therefore, the decisions on the shop floor must take into account the immediate shop floor situation, and eliminate any situation that increases costs. Production scheduling problem is one of allocation of tasks while treating routing as a constraint. Routing is the output of the process planner's decisions. It is a decision taken at too early a stage of the production process. The routing hides the process planner's intentions. Thereby it imposes an artificial constraint on scheduling. The constraints are not real technological constraints; they exist only because of the method used. With a different method, such as by using the process planner table, different constraints or no constraints will exist.

Shop floor control's task is to meet the production plan while treating routing as a variable. It may use the process planner's intentions as well as formal decisions and thus incorporates technology in the dispatching stage, thereby adding a new degree of freedom, and allows crucial decisions to be made at the right time. Hence the foreman's freedom will not reduce processing efficiency, but rather reduce costs.

Mr. F: The Production manager introduced a method that uses the process planning table as a flexible and efficient production planning method. It imitates the RPQ used by purchasing, where each resource competes with others on jobs. If it is so efficient can a similar method used to control shop floor activities?

Mr. PM: Although there are differences between production planning and shop floor control I guess that it might be possible to use the process planning table to control shop floor.

The term *operation* has a different meaning in production management and scheduling and in shop floor control.

Technological operation is an individual processing operation. For example: a process plan might include several technical operations: rough milling, finish milling, center drill, core drill, etc. Each one is regarded as an operation from technological standpoint. The technical operations are the basis for the process planning table.

Production management operation considers an operation as a set of all the activities done on one resource, from loading till unloading. It does not give any indication what the technological operations are.

Production management operations - the routing - are used for production planning and scheduling, while the technological operations are used for machine set up, and preparing work instructions and shop floor control.

Job shop scheduling is generally carried out by means of dispatching decisions, (i.e. at the time a resource becomes free, and a decision must be made regarding what it should do next). Most researches find the SPT dispatching rule (short process time), which means that high priority is given to the waiting operation with the shortest imminent operation time. However, it is up to the process planner to specify a routing with many short processing times instead of few long processing times. The technological operations generated by the process planning table are the shortest process time possible.

Another rule is SIMSET - (Similar Setup), which means that the highest priority is given to the waiting operation with the shortest setup time. This obviously is the best rule for small batch sizes, as it increases machine uptime substantially. The foreman at the shop floor might use this rule informally. However, in the office, production management personnel or researchers cannot use this rule, as conventional routing specifies only the time needed to perform the set-up, and does not specify what has to be done in this time. The detailed information of how to setup a machine for a job is available to the technical routing; it instructs the shop technicians how to prepare the machine for the job. This information is in the technical routing data and is not transferred to the production management stages of the manufacturing cycle.

The process planning table lists all technological operations and their required set up, therefore the SIMSET rule can be used formally on the shop floor by the foreman.

Mr. CC: Following our discussions, the tools such as the process planning table, product structure, released jobs, and data collection equipment that were described may form a close shop floor control system.

The objective of shop floor planning and control is to make sure that the released jobs for a period will be completed on time and in the most economical way possible. Then let it have total flexibility. Instead of making decisions and imposing them on shop floor control, the scheduling should be by objectives. Keep the process planning table, but defer the decision of which path to take to on line control at real time. Furthermore, the decision of a path can be changed after each technological operation. Shop floor control objective is to employ the routing that will result in meeting planned product mix with no bottlenecks or disruptions and at the least operation cost. The objective is not the allocation of tasks.

Mr. FM: Some of what you describe is what I am doing informally. You always refer to my task as meeting due dates; however, I have another responsibility, which is to supply sufficient work to each work station in my department. This is my tool to meet due dates, and keep minimum level of work-in-process. With what you envision my job can be much easier and more efficient. I will change the operation method; instead of allocating jobs to resources, to resources searching for job.

Mr. PM: I would like to summarize the discussion with a proposal for shop floor control based on the ideas presented.

The shop floor control algorithm will be based on the concept that whenever resource is free, it searches for a free job to perform.

A free resource is defined as a resource that just finished an operation and the part was removed, or is idle and can be loaded at any instant.

A free operation is defined as an operation that can be loaded for processing at any instant. An example would be the first operation of an item which the raw material and all the auxiliary jobs (such as: raw material, tools, fixtures, job and quality control instructions, etc.) are done, and is within reach of the machine operator. An intermittent operation is one for which the previous operation has been completed and the part has been unloaded, and is within reach of the required machine.

The method uses the following terminology.

The foreman scans all resources and checks if a resource has become idle. Loading priority is given to the next operation (technological operation) of the same item if it is economical to do so. This is decided by examining the process planning table of that item following the resource column.

If the following operation does not conform to these requirements, or if the resource could not find an appropriate job and was idle for the previous sequential cycle, then the system scans the process planning tables of all free operation items in this particular resource column and makes a list of all free operations that the specific resource can do best.

If the list contains only one entry, then this entry (operation) is allocated to the resource. If the list contains more than one entry, then the system allocates the operation with the biggest time gap of performing it on another resource. This value is determined by scanning the operation row in the relevant process planning tables, and computing the processing time difference between the best resource and the processing time on different resources. Each free operation will be tagged by this difference value. The free operation with the highest tag value will be the one that will be allocated on this sequence cycle on the idle resource.

Table 11-1 demonstrates this algorithm. In this example (R-P) is the idle resource and there are four free operations for which this resource is the best one. The system scans these operations across all resources and computes the difference between the minimum time (R-P) and the time on each resource. The maximum difference value is on the last column marked by **Δ.** In this case the difference between the (R-P) resource and the resource processing time of item 9 operation 2 is the biggest (13.9 - 9.95 = 3.95). Therefore, this operation will be allocated to the (R-P) resource.

Table 11-1. Demonstration of operation selection

Free Operation.	R1	R2	R3	R-P	R7	Rn	Δ
Item 5 Op .4	6.38	6.12	7.05	5.78	5.93	6.83	1.27
Item 9 Op 2	13.9	10.24	10.86	9.95	12.46	999	3.95
Item 3 Op 9	8.34	8.92	7.58	7.23	8.76	8.12	1.69
Item 11 Op 7	11.50	12.87	11.94	11.23	13.12	999	1.89

If the list is empty a "look ahead" feature is used to determine the "waiting time" for such an operation till it will become "free". This search is done by scanning the idle resource column for each one of the items with a search for a best operation. Then the system searches for a free operation that the idle resource can perform economically, although not being the best resource for the job. One method to compute the economics of using an alternate resource is to compute the difference in time between the "best" and the alternate operation. Loading decision is made by comparing the waiting time to the alternate time difference.

As an example, suppose that the quantity of 100 units, the best time is 5 minutes and the waiting time is 150 minutes. The alternate resource processing time is 6 minutes. Then the economic consideration is as follows: producing the operation with the best resource will take 5 x 100 = 500 minutes; producing the operation with the alternate resource will take 6 x 100 = 600 minutes, part of which (150 minutes) replaces the waiting time. Therefore the actual processing time is 600-150 = 450. Hence using the alternate resource and working "inefficient" will save 500 - 450 = 50 minutes of elapsed time.

The search for such an alternative free operation proceeds as follows: The system scans the free operation rows and computes the difference between the best time and the processing time on the idle resource, computing the minimum difference in processing time. The operation with the minimum value will be loaded.

Table 11-2 demonstrates this algorithm. R-idle is the idle resource and there are four free operations. This resource is not the best one for any of the free operations. The system scans these operations across all resources and computes the difference between the processing time of the idle resource and processing time on the best resource. The minimum difference value is on the last column. In this case the difference between the processing time on the best resource and the resource processing time on the idle resource is for item 7 operation 6 is the minimum (5.93 - 5.78 = 0.15). Therefore, this operation will be allocated to the idle resource.

Table 11-2. Demonstration of selecting alternative operation

Free Operation	R1	R2	R3	R4	R-idle	Rn	Min.
Item 7 Op. 6	6.38	6.12	7.05	5.78	5.93	6.83	0.15
Item 11 Op.4	13.9	10.24	10.86	11.33	12.46	999	2.22
Item 17 Op.7	8.34	8.92	7.85	8.23	8.77	9.12	0.92
Item 11 Op.2	11.50	12.87	11.94	11.93	13.12	999	1.62

Mr. F: This algorithm, as I understand it, will meet the objective of minimal idle time of resources. It might meet the objective of minimal manufacturing lead time. But I am not sure that it will keep the capital tied down in production to a minimum, and that it will meet delivery dates. It actually calls to work inefficiently in many cases, is it right to do so?

Mr. C: I understand your doubt, but actually as discussed before, you cannot realize all objectives. But there is a interference between them, what I mean is that if you keep resources idle time to a minimum, then automatically you have increased the working time and thus can increase production and have a better chance to meet delivery dates and decrease capital tied down in production.

I agree with you that working inefficiently lets one understand that costs are going to go up. But what I understood is that the objective is to work inefficiently from a single operation stand point, but being very efficient in processing a product mix. Actually it means to have efficiency by being inefficient. I understand that there are economical considerations to guard and ensure the cost efficiency in the algorithm. I suggest that you and I will be involved in defining the economic algorithms.

Mr. CC: The proposed algorithm seems clear, and with the data of product structure, process planning table, released jobs to shop floor it is easy to develop a computer program that performs as we described.

The president was impressed by the proposed method: "If I understand correctly, it is an extension of the proposed method for production planning. I ask Mr. CC to look into such a computer program and Mr. F to evaluate the economic considerations."

Chapter 12

DECISION SUPPORT

1. INTRODUCTION

The president opened this session reminding those present that the objective is to work as a team and supplement each other, and not to argue about who contributes more to company success. Our group deals with production technology. The topic of this session is management contribution to cost reduction. This topic was proposed by production manager director, and let him now explain what he means by that.

Mr. PM: We had a good discussion regarding cost reduction by enhancing production planning methods. However, in all proposals and methods we were constrained by management decisions. I propose that in this session we will elaborate on this issue.

Manufacturing system is basically an engineering system. It can assist management by supplying information and simulations needed to make decisions of an engineering nature, such as resource planning, expansion of the manufacturing capabilities, and introduction of new manufacturing technologies.

One of the main constraints is the available resources which constrain process planning and thus processing efficiency. Process planning is an important link in the manufacturing cycle. It defines in detail the process that transforms raw material into the desired form. More precisely, process planning defines the operations, sequence of operations, facilities for each operation, and operation details.

Our conclusion to use the process planning table method (PPTM) eased up the constraint by introducing flexibility, but it does not remove this constraint entirely. Processing efficiency establishes the plant level of performance and thus the ability to compete on the market. Processing efficiency is not the only parameter that affects competitiveness, but it is the only one that is relative to our discussion. A company that its resources are better suited to a product mix, have an edge over all other manufacturers.

Resource planning is a management task, and thus it constrains the ability of the engineering production planning stages to achieve full competitive capabilities.

The president rejects the accusation made regarding resource planning. Management of an enterprise is overwhelmingly based on economic considerations. Managing of a company calls for many economic decisions such as capital investment, product line product mix, and resource planning and purchasing. Management regards resource as one of the crucial decisions which it has to make. Management is aware that processing efficiency establishes plant level performance and thus the ability to compete. In order to make a sound decision for resource replacement management relies on economic models and techniques (e.g., total value analysis, ROI, etc.) management turns to the engineering stages with a request for the data that drive the economic models. The engineering recommendations (and economic models) are the basis for the decisions of which resource to purchase is made.

Mr. PM: The engineering data that are fed into the economic model, are those that being asked by management, and not necessarily the ones that will lead to an efficient decision. The need to make investments in resources frequently arises in order to replace old resources. The life of a resource is estimated as 10-20 years. This means that 5-10% of resources have to be replaced every year. New resources usually possess more capabilities than the old ones. Merely replacing resource numbers in the routing file will result in inefficient manufacturing methods. A sound economic decision made in the past, might not be a sound decision at present, in view of the changes and modifications occurred. If optimum processes are to be used, all company routings should be examined and new process plans prepared. However, it is impractical, by using today's techniques, to prepare a new set of process plans whenever a resource is added to the plant. It is a huge job and seldom done in general practice. Processes that might benefit remain unchanged. Thus the data fed to the economic models are incomplete.

Mr. C: An organization for operation is continually undergoing modification and changes in the product mix and quantities of manufactured products. New resources and technologies are introduced and developed. That means that sound decisions made in the past are deteriorating in time and in order to remain competitive a periodical evaluation of the resource efficiency must be made, examining our competitiveness compare to the other companies on the market or the present available technology.

Mr. PP: Such an evaluation requires much expense and work, which probably would not be economically justified. Therefore, it will be done only in cases of crisis (or value analysis) and on a limited scale.

The trend in resource development is toward computerized high-power resources and toward machining centers. The new resources are better qualified and more efficient, but their price is accordingly high. There is no doubt that employing such modern resources may save setup times, increase uptime and quality, reduces material handling, and simplified production planning. However, it is questionable whether they reduce production costs. In many cases a 35 KW machine with 5 degrees of freedom, that costs about $800,000 is employed in drilling a series of ¼" holes. Such an operation may be carried out more efficiently, by a $1000 drill press. In the metal cutting process a rough cut usually preceeds finish cut. A rough cut (in metal cutting) does not require accuracy and may be produced by an old inaccurate resource, which probably was fully depreciated. Employing modern resources for all operations no doubt will reduce manufacturing time and result in ease of managing. However, it will not always result in the minimum production cost. Therefore, re-evaluation of process planning of all products should be made and supply to management to make the decision.

Mr. CC: The process planning table method (PPTM), as was demonstrated in chapter 9 (table 9-1), solves this problem automatically. The computer program selects the sequence of operations and which resource will perform each operation. The automatic resources probably will be selected for maximum production optimization, but not for minimum cost optimization. If we based our production planning, stock allocation, production scheduling, job released, and shop floor control on the PPTM than the same tool can be used for evaluating the compatible of resources to the product mix, and it can be done by a computer program in a very short time.

Mr. C: I do not understand how the process planning table method can evaluate our competitiveness to other manufacturers. To do so we must have the routings of all other manufacturers, and I doubt if they will supply such data to us. PPTM was excellent for our production planning tasks, but not for evaluation.

Mr. CC: PPTM was partially described before by showing table 9-1 and its uses. But it did not explain how such table was constructed, or can be constructed. Let me do it now.

PPTM is part of a Computer Aided Process Planning (CAPP) program. The CAPP program is composed of three stages:

- Technology,
- Transformation,
- Decision (mathematics).

The Technology stage generates a TP - Theoretical Process. The Transformation stage constructs the table. The Decision (mathematical) stage solves the table and generates a dynamic process plan according to the immediate shop dynamic requirements.
A Theoretical Process is a fixed universal reference point. Its value is based on actual available technology. It considers only technological constraints. It assumes an imaginary resource; that is, no resource constraints are considered. Thus, the TP process plan is practical from an engineering standpoint and theoretical from a specific shop standpoint. Its value does not include set-up cost. Consequently, it is free from sales, lot sizing, grouping, and scheduling effects. It is a theoretical value that most probably will never be achieved. However, it is a fixed value, representing the state of technology. The numerical value of the TP is expressed in time or cost units. The conversion from time to cost is accomplished by multiplying the processing time by hourly rate. The hourly rate for the imaginary machine can be set as the lowest hourly rate used. This guarantees that the dispersion will be to only one side of the fixed reference point.

The transformation stage constructs the process planning table. The equations for transformation are straightforward; a computer program can easily be developed to perform this task.

The content of the table is $T_{i,j}$ which is the time to perform each theoretical operation (i) on each one of the practical candidate resources (j). The theoretical operations (TP) are translated and adjusted to comply with each individual resource features. It is obvious that the machining time cannot be decreased, it may only be increased. The adjustment considers the following factors: resource physical size, accuracy, special features, available power and torque, available speeds and feeds, number of tools, type of controls, handling time etc. A resource file containing resource specifications is used for the conversion.

Solving the PPTM is a Practical Process - PP; it is a fixed specific shop reference point. Its value is based on the available resources in a specific shop. The PP is practical from the standpoints of technology and available facilities and theoretical with regard to production and capacity planning, that is, the availability of the required machine at the required time.

2. RESOURCE LEVEL OF COMPETITIVENESS

Mr. PM: Management is responsible for the available company resources; the role of engineering is merely a consultant one. When a resource is candidate for replacement, merely replacing it by another resource that performs the same operations will result in inefficient manufacturing methods. New resources call for re-evaluation of all routings. However, it is a huge job and seldom done in general practice. Processes that might benefit remain unchanged. Thus the data fed to the economic models are incomplete.

The proposal of the PPTM method can be used for the evaluation of all routings in a very short period of time and be economical. This is due the fact that with PPTM there is no fixed routing; routings are not stored, but recomputed every time they're needed. What is stored are a list of operations for each item, and a list of all available resources (at the time of generating a routing) and the computer program which devises a routing.

The resource level of competitiveness is defined as the suitability of the available resources to the company product mix. A company that has the most suitable resources has an edge in competing in the market. The level of competitiveness is defined as a machinability ratio - MR. MR is measures on a scale from zero to one, where one is the most suitable resources.

The ratio of time (cost) to produce an item with the existing resources (PP) to time (cost) to produce the part by existing technology (TP) establishes the MR is:

$$MR = TP/PP \tag{1}$$

Mr. PP: Can PP be supplied by a process planner and not by the PPTM system?

Mr. PM: No; PP must be computed by the PPTM system. Remember that the method measures company efficiency and not specific process planner expertise. To rank process planner level of expertise we may compare his routing (MPP - manual process planning) to PP:

$$\text{Process planner rank} = MPP/PP \tag{2}$$

Mr. PS: I do not understand how this equation can establish the company level of competitiveness, as it is based on only one item. In our company any resource is used to process many items, the computation must include all or several items.

Mr. CC: The machinability rating for several items may follow the logic of the single item resource rating equation (1). The rating is the sum of single item MR divided by the number of parts, and will have the following form:

$$MR = \frac{1}{p}\sum_{i=1}^{p}\frac{TP_i}{PP_i} \quad (3)$$

Where MR = machinability rating, p = number of items, PP_i and TP_i = practical and theoretical processing time of item i.

Mr. PS: Equation 3 is an improvement, but it ignores the quantity effect. It does not make sense that all items are produced with the same quantity. Can the equation be further improved?

Mr. CC: You are right, to consider the quantities of each item, a modification to equation 3 is made. MR is replaced by MRQ, were each individual item gets a weight according to its quantity. To arrive at MRQ for an individual item, its MR is multiplied by its quantity. The sum of the individual MRQs is divided by the total quantity to arrive at the company MRQ value. Thus, the equation is:

$$MRQ = \frac{1}{\sum_{i=1}^{p} Q_i}\sum_{i=1}^{p}\frac{TP_i}{PP_i}\times Q_i \quad (4)$$

MRQ represents the routing efficiency of product mix of p items (not products).

Mr. F: I am not sure that the averages have any practical meaning. More important is to know the effect of each order, or each item separately. Such data may be used to determine which items are best suited to our available resources, which are not. Furthermore, the company might have the best resources for producing items, considering the quantities, yet its competitiveness might be jeopardized by resource idleness, i.e. not having enough loads to keep it occupied. To arrive at overall optimization, management must have a load profile and data on resource utilization.

Mr. CC: At your request we will define machine utilization rating - MUR. MUR is defined as the load rating. Single resource load rating is defined as the resource load to the "maximum load resource" value. Load is defined as the processing time on the resource. This definition is best explained by an example.

Table 12-1. Routine for part "sample"

Operation	Resource	cost	time
010	4	1.62	1.62
020	1	2.13	0.71
030	5	3.52	1.76
040	2	10.32	7.37
Total		17.59	11.46

Table 12-1 gives the PP process, it calls for four operations and four resources, the processing time and cost of each operation is specified. The load on resource 2 is the greatest; hence it is the "maximum load resource". Therefore MUR on resource 4 is 1.62/7.37= 0.22; resource 1 is 0.71/7.37=0.096; resource 5 is 1.76/7.37=0.239; and resource 2 7.37/7.37=1.00. The average load (if it has any practical meaning is (0.22+0.096+0.239+1.0)/4=0.389.

Mr. PM: Let have an example of the method, and follow Mr. F note; the most useful data to management is the rating of individual item, and individual resource. Table 12-2 demonstrates an example of computing item and resource rating, for individual and the average.

Table 12-2. Example of computing MR and MUR

Item	TP	R1	R2	R3	R4	R5	R6	PP	MR
1	7.37	0.71	7.37		1.62	1.76		11.5	0.67
2	12.3	2.28	9.4	5.93	4.78	5.6		28.0	0.44
3	4.80		3.21	2.83	3.15	1.24		10.4	0.46
4	17.5	6.25	8.65	9.64	6.38		12.5	43.4	0.40
5	13.5		4.03	7.83	4.38		1.51	17.7	0.76
6	14.9	3.89		11.1	8.76	4.58	8.05	36.4	0.41
Σ	time	13.1	32.7	**37.3**	29.0	13.2	22.1	Σ	3.14
MUR		0.35	0.87	1.00	0.78	0.35	0.59	MR	0.52

These results can be presented to management as a diagram, as shown in Fig. 12-1.

Engineering are not economist experts; their task is to supply data to management, as the one proposed here, at let management make decisions. The decision might be to change the product line, push sales of slow moving products, or purchase new manufacturing resources.

Mr. F: The order quantity has an effect on the processing time and thus on resource utilization time, but I notice that it was not taken into consideration in the rating.

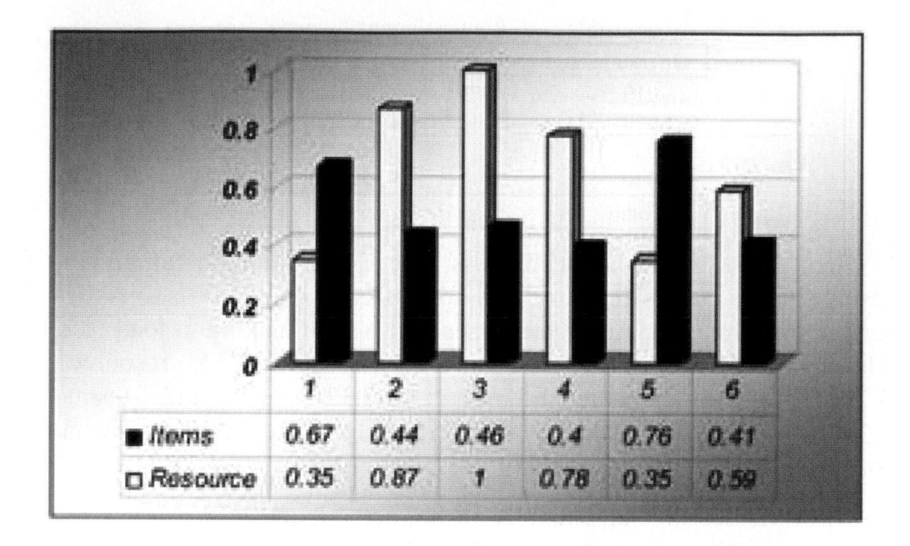

Figure 12-1. Items and resource rating

Mr. PM: the quantity is not an engineering data, and it is a variable. The proposed example is the framework of computation. Management may add the quantity to the equation and the table, and then compute the utilization rating.

3. RESOURCE PLANNING

Mr. PR: When the need to purchase a new resource arises, a list of alternate resources is assembled, usually based on catalogs, vendor information, and specification of old resources, or random choices. We issue to the candidate a request for quotation. The quotations are returned to the process planner for evaluation.

Mr. PP: We evaluate the proposals, generate a process plan for each resource and transfer recommendations to management for economic decision, and then back to purchasing to negotiate terms with the selected supplier.

It would take quit a lot of time, cost and effort to evaluate all the alternate resources, and the effort would probably not be economical. Therefore, the process planner proposes a limited number of alternatives (if at all) and let the economist decide which one of them to select. Hence, the "best" alternative might not be even being considered, and a biased decision might be reached.

Mr. PR: We are at execution stage; our objectives are to obtain the required items and resources, at the required quantity and quality at the right time. It is not for us to question why the need is specified. Our decisions are concerned with selecting a supplier subject to the optimization criteria of quality, quantity, delivery date, and cost.

Mr. PS: We have to work with the given resources, routings. Once the process planner makes a decision, it becomes a constraint. An artificial constraint; they are in effect only because of the sequence of decisions made. Another decision might result a different set of constraints, and therefore results in a different schedule.

Mr. C: The method described is quite discouraging; each stage is doing his task efficiently, but the total system suffers from artificial constraints that prohibit the competitive effort. How can we eliminate artificial constraints?

Mr. CC: The concept of PPTM that we discussed before was developed to overcome the problem of artificial constraints. By this concept a process is generated using only real constraints. By employing the theoretical process - TP concept, the process planner generates a process plan in the usual way, but using an imaginary resource.

The TP process is theoretical from a specific shop viewpoint, but it is practical from a technological standpoint. It does not violate any physical or technological rule. In this sense the TP indicates the most desirable resource characteristics and features. The term "imaginary resource" might be ambiguous and frightening. It is a resource with unlimited power, with infinite speed etc. However, one does not have to be alarmed. The "imaginary resource" is the resource that possesses the requirement specifications to perform the TP process plan. Several operations are required to produce a part. There are rough operations that require heavy forces and limited accuracy, while finishing operations require light forces but a significant accuracy. The process considers many real constraints such as part specifications, part shape and strength, fixture etc. Therefore most operations will require commercially available resources, and only few operations might require special resources.

Each operation specifies the power, moment, forces, speed, revolutions per minute, feed rate, size of part, the accuracy required by the operation etc. These data are actually points to the "best" characteristics that a resource should possess, in order to perform the particular operation in the most economical way.

Therefore, the needs of the individual TP operation will be used as a specification for RFQ - Request for Quotation that will be distributed to suppliers.

Mr. PP: To evaluate the proposals, I may use the PPTM method to generate a routing, but instead of using available resources the RFQ proposals are used as the available resources.

At this stage the characteristics of each individual resource are known by the quotation received. Adjustment of the theoretical operation for each individual resource can take place and the process planning table is ready to generate routings. By using PPTM the solution is transferred from a technological problem to a mathematical one. This method can be used to generate alternative routings. An alternative routing is generated by ignoring a resource from the table and re-computing a routing. The financial planner can generate as many alternatives as he or she desires. Such alternatives can be put into a spreadsheet to compute the optimum investment according to company policy.

Mr. F: The economic model may vary from one plant to another. However, the basic data that goes into the model are similar. The required general data might include: machine cost, finance cost, installation cost, maintenance cost, energy consumption cost, labor cost, life cycle, etc. These data are available from the quotation supplied by the machine manufacturer and plant economic accumulated experience. The required technical data include the machining time per part, the cost of machining a part can be furnished by the process planning table.

In resource planning application, the target is to evaluate cost - performance of alternative resources. The role of the process planning table is to supply objective data to management, who will make the decision. To accomplish this task the computer program is programmed to generate many alternative processes, using different resources, and different criteria of optimization, different lot sizes, and penalties. The purpose of generating the alternatives is to prepare data that reflect machining time and cost as a function of the investment in purchasing a new resource.

For demonstration purposes assume that the RFQ proposed six resources, their purchasing cost is specified, and assume that processing hourly rate is proportional to the purchasing cost.

The assumed relative purchasing cost is as follows:

RFQ#1	RFQ#2	RFQ#3	RFQ#4	RFQ#5	RFQ#6
1.0	0.5	1.5	1.3	0.7	0.3

The process planning table generated 10 alternatives as shown in table 12-3

Table 12-3. Alternative resources

Alternative	Resources	Total time	Total cost	Relative investment	Coefficient of investment
1	5	9.64	6.75	0.7	233%
2	6	32.32	9.70	0.3	**100%**
3	3; 5	**7.66**	8.00	2.2	733%
4	2; 3; 5	8.50	7.45	2.7	900%
5	5; 3	9.09	8.18	2.2	733%
6	6; 3; 5	12.45	8.32	2.5	833%
7	2; 5	8.65	**5.39**	1.2	400%
8	5; 2	9.74	6.61	1.2	400%
9	6; 2; 5	13.10	6.76	1.5	500%
10	3; 2; 5	8.81	6.93	2.7	900%

The "Relative investment" gives the relative cost of the resource RFQ that has to be purchased. If more than one resource RFQ is used in the alternative then the sum of the relative cost is given. For example alternative 1 requires only resource 5 whose relative cost is 0.7. Alternative 4 requires the use of resource 2; 3; 5 therefore its relative cost is the sum of these three machines 0.5+1.5 + 0.7 = 2.7.

The "coefficient of investment" column is the cost relative to the minimum cost of investment. The smallest relative machine cost is that of resource 6 which is 0.3, and is regarded as the 100% investment. All other alternative "coefficient of investments" are computed relative to this minimum value. Hence alternative 2 will be 100%, the minimum required investment. While the investment for alternative 4 (or 10) is (2.7/0.3) x 100 = 900% meaning 9 times that of the minimum investment.

The effect of the amount of investment on machining time and cost, is shown in Fig. 12-2 sorted by investment cost.

Mr. C: Examining the data in table 12-3 and the Fig. 12-2 reveals some astonishing facts; increase in investment by no means assures better optimum (alternative 7). Furthermore, it indicates that the "best" machines for the maximum production criterion of optimization (alternative 3) are not the same as for minimum cost criterion.

The data clearly indicates that there is no direct correlation between the investment and optimum process plans. Alternative 2 is the lowest relative investment and the process the gives the worst machining time, four times longer than the optimum machining time (32.32/7.66 = 4.22); however, it may result in one of the two best investments from the ROI standpoint.

Mr. CC: It seems that management will have a challenging task in deciding which resources to purchase. Comparing the minimum time (alternative 3) to the minimum cost (alternative 7) indicates that increasing the machining time by 13% (from 7.66 minutes to 8.65 minutes) will reduce machining cost by 67%, (from $8.0 to $5.39) and the investment may be reduced by 55% (from 733% to 400%).

The decision as to which resource to purchase must consider many parameters. The PPTM method is not intended to make an economic decision; its sole purpose is to supply sound engineering data to the decision makers.

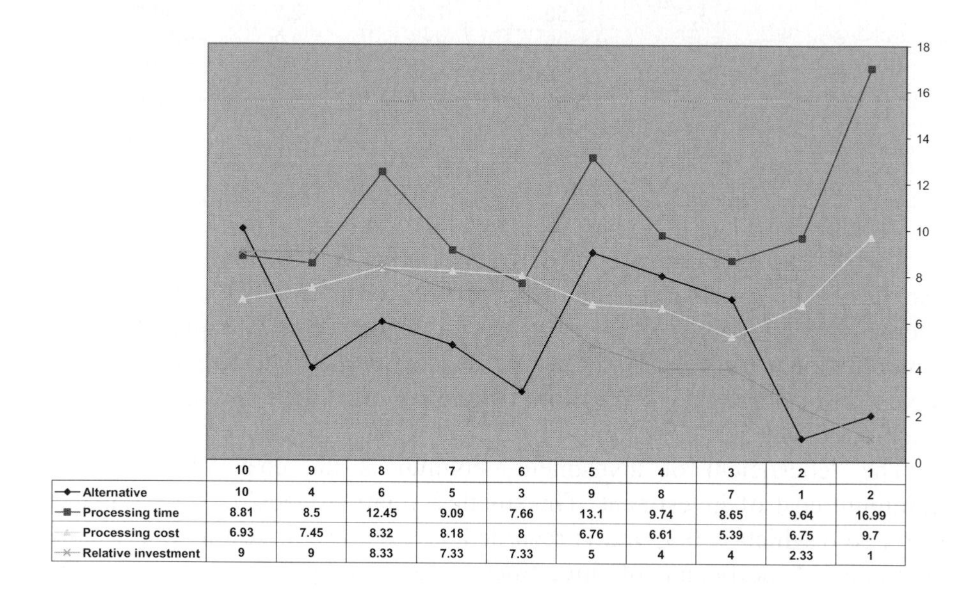

	10	9	8	7	6	5	4	3	2	1
Alternative	10	4	6	5	3	9	8	7	1	2
Processing time	8.81	8.5	12.45	9.09	7.66	13.1	9.74	8.65	9.64	16.99
Processing cost	6.93	7.45	8.32	8.18	8	6.76	6.61	5.39	6.75	9.7
Relative investment	9	9	8.33	7.33	7.33	5	4	4	2.33	1

Figure 12-2. Relationship of Investment to machining time and cost

Mr. PM: The utilization time of each resource can be crucial in making a decision. This information is also immediate available from the solution of the PPTM. Naturally, this figure is a function of the quantity required. The PPTM solution handles the unit time and cost. The quantity affects the (penalty) but not the direct machining time. However, the total utilization time per a period can be computed.

Mr. PS: Naturally, the total required quantity has to be taken into consideration. For a very low quantity resource 6 (alternative 2) will probably be preferred. For higher quantity resource 5 (alternative 1) should be preferred. However, if single resource 5 cannot handle the load, then 2 resources 5 are needed. In that case, it is better to purchase one resource 5 and one resource 2 thus reducing the investment from (0.7 x 2=) 1.4 to 1.2. The best combination may be decided by examining the data in table 12-3 and figures 12-2.

Mr. FM: It is unlikely that any single item will supply complete load to any resource and balanced the load for several resources. Increasing load and load balancing may be done by considering other items. Therefore, considering many items in one run is preferred. The parts might all be from one new product, or parts already in production.

The president thanks the participants in this session. It is management's responsibility to make resource planning decisions; however the proposed method allows management to work with a computer model that will supply engineering data, in any desired form, instead of calling the engineer each time information is needed. Different plants will use different economic models. Therefore, we regard the proposed method as a data generator and not as a recommended mode. The method by which the table presents the data may vary from one plant to another. Additional data, if required, may also be retrieved from the PPTM format method.

Appendix – 1

STATISTICAL PROCESS CONTROL

1. INTRODUCTION

Statistical Process Control (SPC) is a technique for error prevention rather than error detection. Products of the required quality will be produced not because they are inspected but rather because they are manufactured properly.

The goals of an SPC program are consistent with typical company goals of:

- Improve quality
- Reduce manufacturing cost
- Increase profit
- Enhancement of a competitive advantage

SPC analyzes and controls the performance of the activities performed on the manufacturing processes. A controlled process offers many advantages to both the producer and the consumer.

The producer will attain lower production costs, lower rework and scrap costs. In addition to these economic considerations, effective process control may justify the reduction in the amount of inspection and test performed on final products.

Specific goals of SPC are as followed:

1. To improve quality and reliability of products without increasing cost. This objective is not simply an intrinsically 'good' thing to do, but is a necessity for an organization that wants to remain competitive. Steps

taken to improve a process will result in fewer defects and therefore a better quality product delivered to the consumer.

2. To increase productivity and reduce costs Application of SPC can produce immediate improvements in yield, reduce defects and increase efficiency, all of which are directly related to cost reduction.
3. To provide a practical working tool for directing and controlling an operation or process. Implementation of SPC creates a high degree of visibility of process performance. The same statistical technique used to control the process can be used to determine its capabilities.
4. To establish an on-going measurement and verification system. Measurements will provide a comparison of performance to target objectives and will assess the effectiveness of problem solutions.
5. To prioritize problem-solving activities and help with decisions on allocation of resources for the best return on investment. SPC directs efforts in a systematic and disciplined approach in identifying real problems. Less time and effort will be spent trying to correct non-existent or irrelevant conditions.
6. To improve customer satisfaction through better quality and reliability and better performance to schedule.

Effective process control will enable the producer to fulfill his responsibility of delivering only conforming products and on time. When an out-of-control process is detected, the producer should attempt to identify and correct the cause since the nonconforming products being produced are probably not due to chance.

The consumer receives products of the required quality and on schedule. Additional benefits are: improved quality which will result in lower maintenance, repair, and replacement cost, less inventory of spare parts, higher reliability, better performance and reduced time lost due to defective products.

Benefits of SPC include defect or error prevention rather than just merely detection. This means more machine up-time, less warranty costs, the avoidance of unnecessary capital expenditures on new machines, an increased ability to meet cost targets and production schedules and increase productivity and quality.

Additionally, SPC has been used as a basis for product and process design. With detailed knowledge obtained from SPC on product variability with process changes, designers have the capability to design and produce items of the required quality from the first piece. Therefore, SPC control not only helps with design, but results in reduced start-up and debugging effort and cost.

1.1 Basic SPC tools

SPC is statistically based and logically built around the phenomenon that variation in a product is always present.

There is a natural variation inherent in any process due to wear of tools, material hardness, spindle clearance, jigs and fixtures, clamping, machine resolution, repeatability, machine accuracy, tool holder accuracy, accumulation of tolerances, operator skill etc. Variation will exist within the processes. Parts that conform to specifications are acceptable; parts that do not conform are not acceptable. However, to control the process, reduce variation and ensure that the output continues to meet the expressed requirements, the cause of variation must be identified in the collected data or in the scatter of data. Collection of these data is characterized by a mathematical model called 'distributions' which is used to predict overall performance.

Certain factors may cause variation that cannot be adequately explained by the process distribution. Unless these factors, also called 'assignable causes' are identified and removed, they will continue to affect the process in an unpredictable manner.

A process is said to be in statistical control when the only source of variation is the natural process variation and 'assignable causes' have been removed.

SPC identifies changes between items being produced over a given period, and distinguish between variations due to natural causes and assignable causes. Corrective action may therefore be applied before defective products are produces. A properly conducted SPC program recognizes the importance of quality and the need for never ending search to improve quality by reducing variation in process output. Parts will be of the required quality because they are manufactured properly, not because they are inspected. SPC is basically opposed to methods involving part sorting, such as sorting of conforming parts from nonconforming ones.

Variations that are outside of the desired process distribution can usually be corrected by someone directly connected with the process. For example, a machine set improperly may produce defective parts. The responsibility for corrective or preventive action in this case will belong to the operator, who can adjust the machine to prevent recurring defects. Natural variation will establish process capability. The process must be in control in order to apply SPC. A process in control has its upper and lower control limits, which establish the suitability of the process to the task and the anticipated scrap and rework percentages. Inherent capability of process factor (*Cp*) will indicate if:

- the process is capable;
- the process is capable but should be monitored;
- the process is not capable.

Natural process variation may only be corrected by redesigning the part and the process plan.

Successful SPC control requires action in a form of a monitoring system and feedback loop, in a corrective and preventive action plan. A control chart may be in place to record the average fraction defective at a work station, but it is only of marginal value unless the people responsible for the process know what action to take when the process moves out of control.

SPC eliminates subjectivity and provides a means of comparing performance to clearly defined objectives. The control chart used to identify variability and existence of assignable causes will be used to track process improvements.

Through application of statistical techniques, problems are identified, quantified and solved at the source in an optimum time. Out-of-Control conditions become evident quickly, as does the magnitude of the problem. With this information, action can be taken before the condition becomes a crisis.

Immediate feedback is the key to the success of any SPC system. SPC is not solely a quality department function. The responsibility for control is in the hands of the producer. This provides the dual advantage of giving the operator a better understanding of what is expected, as well as providing a means of detecting undesirable conditions before it is too late.

2. PREREQUISITES FOR SPC - PROCESS CAPABILITY

Process capability is the measure of a process performance. Capability refers to how capable a process is of producing parts that are well within engineering specifications.

A capability study is done to find out if the process is capable of making the required parts, how good is it, or if improvements are needed. It should be done on selected critical dimensions.

To employ SPC and the process capability, the process must be under control and have a normal distribution. If the process is not under control, normal capability indices are not valid, even if they indicate that the process is capable.

There are three statistical tools to use in order to determine that the process is under control and follows a normal distribution.

1. Control charts
2. Visual analysis of a histogram
3. Mathematical analysis tests

The control charts (which will be explained in the next section) are used to identify assignable causes. The capability study should be done only for random variations data.

A histogram is a graphic representation of a frequency distribution. The range of the variable is divided into a number of intervals (usually, for convenience, of equal size) and a calculation is made of the number of observations falling into each interval. It is essentially a bar graph of the results. In many cases a statistical curve is fitted and displayed on top of the histogram.

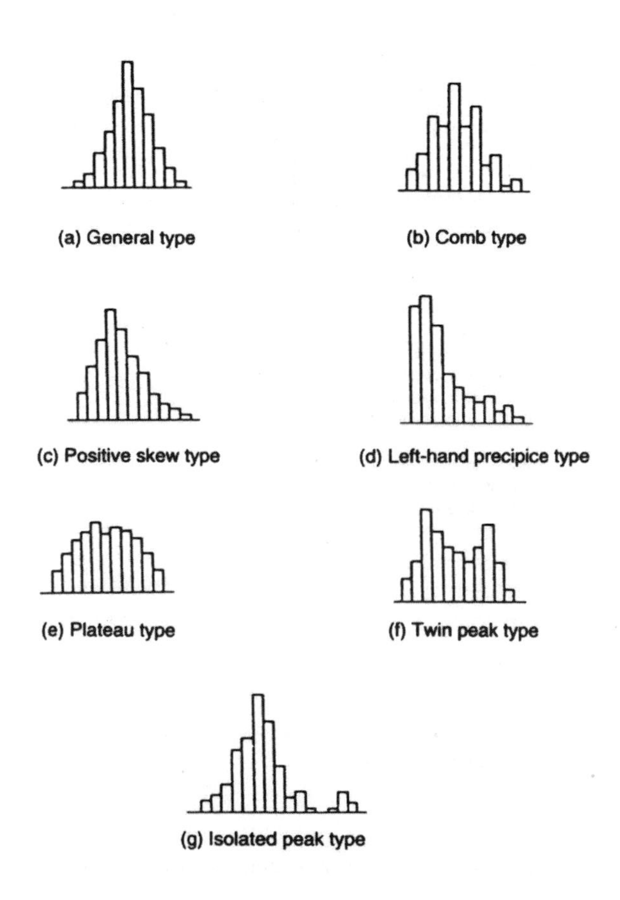

Figure Appendix – 1-1. Types of histograms

It is possible to obtain useful information about the state of a population by looking at the shape of the histogram. Figure Appendix 1-1 shows typical shapes; they can be used as clues for analyzing a process. For example:

Case (a): general type, normal distribution; the mean value in the middle of the range of data and the shape is symmetrical.

Case (c): positive skew type; the process capability may be excellent, i.e. it uses only part of the tolerance. However, problems of excessive variation caused by shift and out-of-control may appear. The cause may be attributed to the machine, the operator, or the gauges.

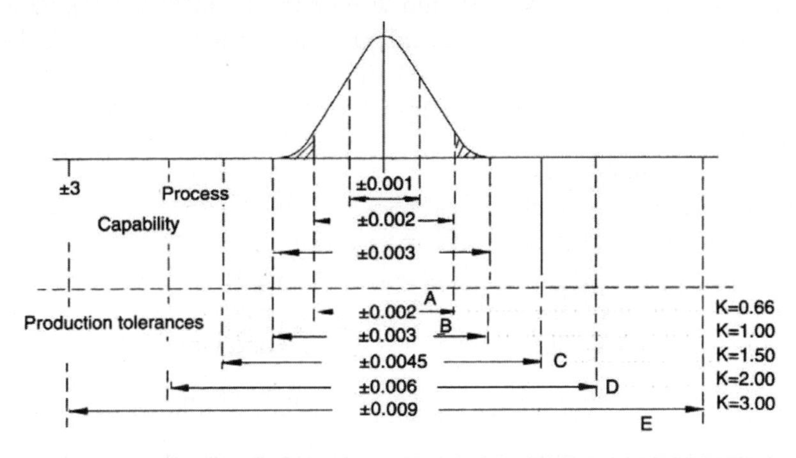

Case	Remarks	Samples
A	High production risk; any shift in average will increase failure	100% inspection is a must
B	No room for process, average shift, accurate setup	1:1 1:3 1:5 depending on part value
C	Standard requirement of system	1:5 1:10 1:15 depending on part value
D	Improve type C	1:15 1:20 1:30
E	Wide open tolerance, no production problems anticipated	Once a day or batch

Figure Appendix – 1-2. Production tolerance vs. process capability

The ±3σ of a normal distribution curve is regarded as a reasonable process capability and can be computed by known equations. The capability of the process to meet engineering specifications is the comparison of the ±3σ with the tolerance. Fig. Appendix 1-2 shows the production tolerance versus ±3σ of process capability. If the tolerances are within the ±3σ, it means that there will be rejected parts. The probability of the percentage of rejects, rework and scrap, can be computed by known statistical method. If the tolerances are much wider than the process capability, no production problems of size are encountered, and probably inspection and SPC is not needed. The most commonly used capability indices are C_P and C_{pk} . C_P standing for Capability of Process is the ratio of tolerance to 6σ. It is computed by:

$$C_P = tolerance / 3\sigma$$

As can be seen, the greater the C_P values the better the process. $C_P = 1$ means that 99.73% of the parts will be within engineering tolerances. However, any minute deviation from the mean will produce more rejected parts. Therefore, it is usual to aim for $C_P = 1.33$. Nowadays, a target of ±6σ or $C_P = 2$ becomes a dominant figure.

C_P is only a measure of the spread of the distribution; it is not a measure of centering. The distribution midpoint may not coincide with the nominal dimension, and thus, even when the C_P shows good capability, parts may be produced out of specification. Therefore, the C_{pk} index is introduced.

C_{pk} is a measure of both dispersion and centeredness. It computes the capability index once for the upper side and then for the lower side and select as the index the lower of both.

3. CONTROL CHARTS

Control charts are the tools for statistical process control. Statistics and parameters by themselves are hard to interpret and visualize. The control chart however is a pictorial method which enables the operator to tell at glance how well the process is controlling the quality of items being produced.

There are essentially two kinds of control charts, control charts for variable data (quantitative measurements) and one for attribute data (qualitative data or count). The variable control charts are more sensitive to changes and therefore are better for process control. The average value - $\overline{x}$ and the range - R chart is the most common form of control charts, and one of the most powerful for tracking and identifying causes and variations.

Since the parts coming off a process are in large numbers, we need a way to establish and monitor the process without having to measure every part. This is done by taking samples (subgroups) of 2 to 10 parts (five as the most common size) and plotting the measurements on a chart, (Fig. Appendix 1-3). The side-way distribution curve (histogram) represents the mean and the range of each subgroup. The $\overline{x}$ and R chart (Fig. Appendix 1-4) is easier to make. Instead of calculating and graphing small histograms of data, separate graphs for the mean ($\overline{x}$) and the range (R) are used.

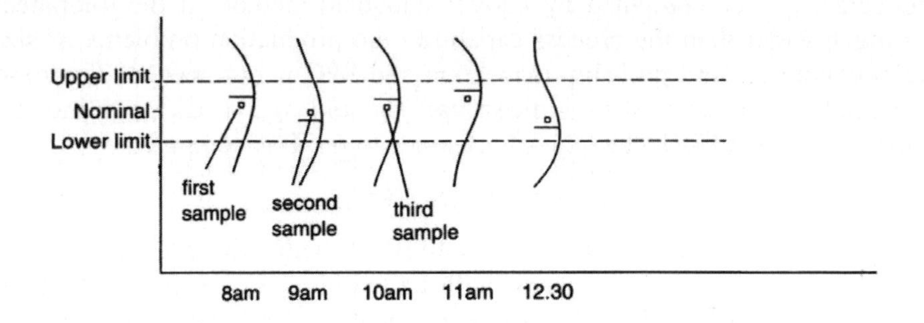

Figure Appendix – 1-3. Subgroup and samples

To interpret the chart at a glance, the centerline and control limits are drawn on the chart.

The centerline is the average of the $\overline{x}$ values.

The purpose of the control limits on the chart is to indicate if the process is under control, i.e. that 99.73% of all the average of subgroups $\overline{x}$, will be within these limits. More accurately, under-control customarily means that all $\overline{x}$ are within the estimated $\pm 3\sigma\overline{x}$ limits of the process.

According to statistics theorems, the sample mean from a normally distributed population is exactly distributed as a normal distribution, and, even if the distribution of a population is not normal, the sample mean is approximately a normal distribution. The approximation holds best for large samples (*n*), but is adequate for a value of *n* as low as 5. The formula to calculate the estimated sample deviation is:

$$S = \sigma / n^{-2}$$

Just as the σ is a measure of variation in a sample, the *s* is a measure of variation that may be expected when obtaining one observation ($\overline{x}$) from the distribution mean. Fig.Appendix 1-5 shows the relationship between σ and s.

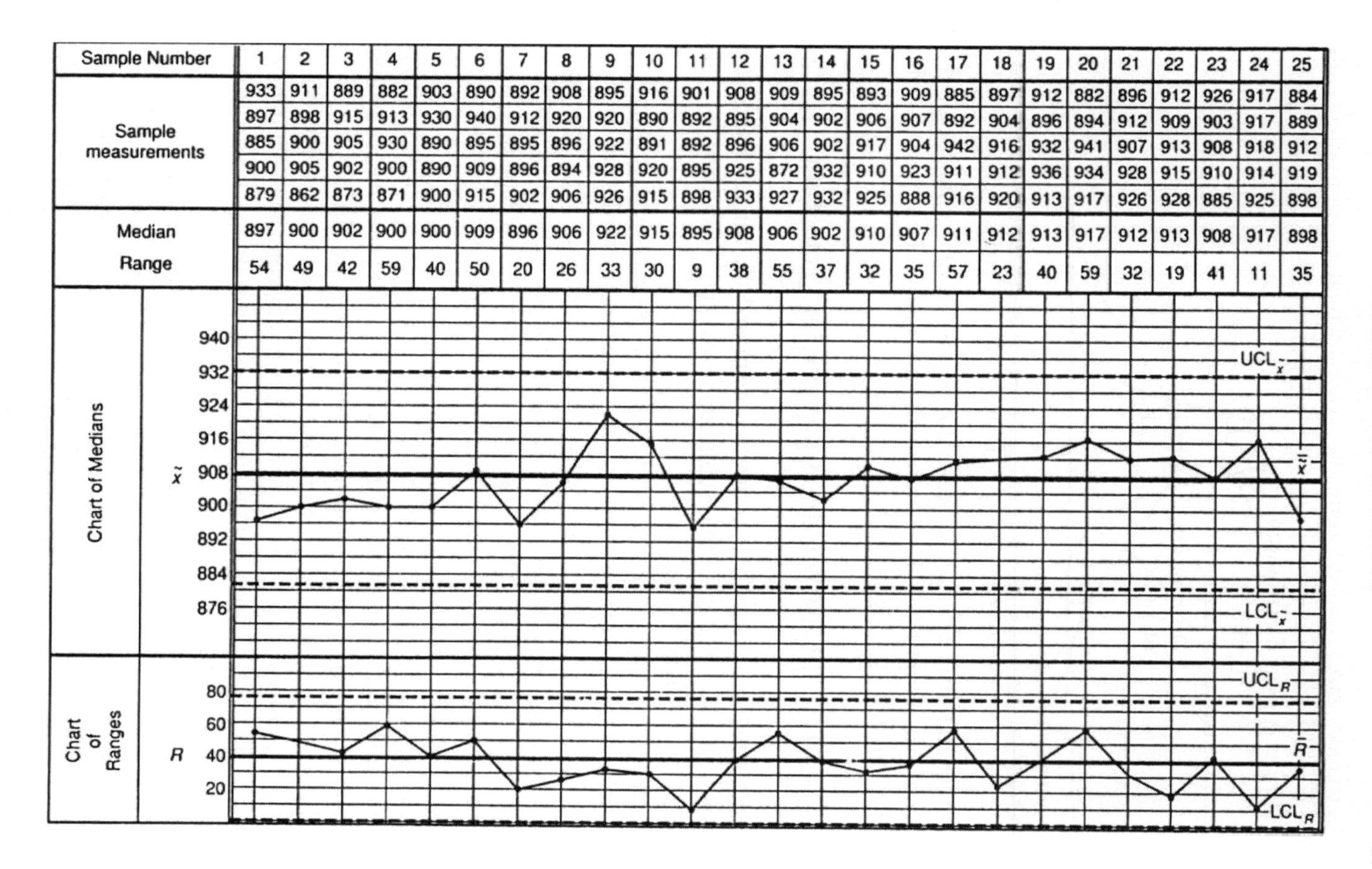

Sample Number	1	2	3	4	5	6	7	8	9	10	11	12	13
Sample measurements	933	911	889	882	903	890	892	908	895	916	901	908	909
	897	898	915	913	930	940	912	920	920	890	892	895	904
	885	900	905	930	890	895	895	896	922	891	892	896	906
	900	905	902	900	890	909	896	894	928	920	895	925	872
	879	862	873	871	900	915	902	906	926	915	898	933	927
Median	897	900	902	900	900	909	896	906	922	915	895	908	906
Range	54	49	42	59	40	50	20	26	33	30	9	38	55

Sample Number	14	15	16	17	18	19	20	21	22	23	24	25
Sample measurements	895	893	909	885	897	912	882	896	912	926	917	884
	902	906	907	892	904	896	894	912	909	903	917	889
	902	917	904	942	916	932	941	907	913	908	918	912
	932	910	923	911	912	936	934	928	915	910	914	919
	932	925	888	916	920	913	917	926	928	885	925	898
Median	902	910	907	911	912	913	917	912	913	908	917	898
Range	37	32	35	57	23	40	59	32	19	41	11	35

Figure Appendix – 1-4. The control chart

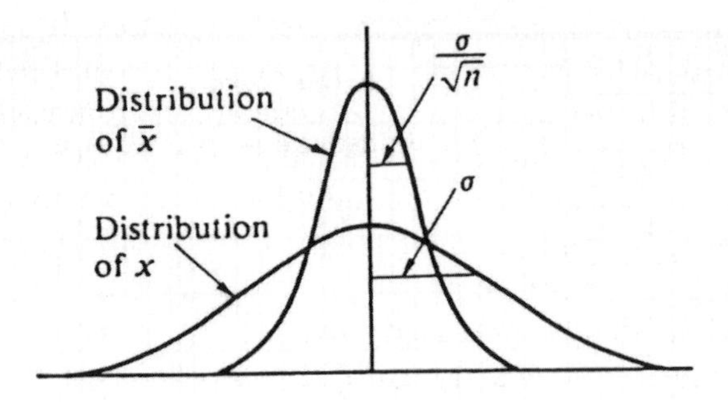

Figure Appendix – 1-5. Distribution of $\overline{X}$ and R

The general rule is that at least 20 points on the control chart (20 subgroups) representing 100 measurements are needed before control limits can be calculated.

3.1 Control chart parameters selection

The points on a control chart are the mean and the range of the sample subgroup. In this section, the subgroup size and frequency of taking measurements will be discussed.

A rational subgroup is one where there is a very low probability of assignable causes creating variations in measurements within the subgroup itself. If a subgroup has five measurements, then the opportunities for variation among those measurements must be made deliberately small. This usually means the subgroup should be taken from a batch of pieces made when the process operates under the same setting - one operator and no tooling or material changes.

Five consecutive pieces might be the easiest to collect. The logic behind rational sub-grouping is that if the variability between pieces within a subgroup is entirely due to common causes, then the differences in sub-grouping averages and range will be due to assignable causes. The effect of assignable causes will not be buried within a subgroup and dampened by averaging. They will appear on the chart in the form of a point that exceeds the control limits or have an identifiable pattern.

If the time of day may contribute to variations between pieces, then one subgroup of five consecutive pieces from the process should be collected at selected times through the day. The time interval between subgroups reflects the expected time of variations in the process and the cost and ease of taking the measurements. In stable processes, every few hours might be satisfactory. In processes where tools wear rapidly, or other changes in short periods of time exist, short intervals should be used.

An $\bar{x}$ and R chart is used to plot one set of causes. If several different machines contribute to a single lot of parts, a chart of samples taken from the lot will not reveal nearly as much as separate charts from each machine. On the other hand, if one machine with the same setup produces different parts, one chart may be sufficient to control all the parts, as we control the process and not the parts. To do it, a chart of deviations from the nominal is used instead of charting the nominal itself.

In cases of small lots produced on the same machine with the same tooling, the technique of charting 'moving averages' or nominal or σ might be the answer.

4. INTERPRETING CONTROL CHART ANALYSIS

Analysis is accomplished by the use of control charts, mainly the R chart and the $\bar{x}$ chart. The most common feature of a process showing stability is the absence of any recognizable pattern. The points on the charts are randomly distributed between the control limits. A rare point out of limits on a process that has shown stability over a long period of time can probably be ignored.

The characteristics of a stable process are:

- Most points are near the centerline;
- Some points are spread out and approach the limits;
- No point beyond the control limits.

The characteristics of non-stable process are:

- points outside the control limits;
- four out of five successive points outside ±σ limits;
- points crowded near centerline;
- two out of three successive points falling outside ±2σ limits on the same half of the chart;
- a trend of increasing or decreasing trend of seven points;
- 12 out of 14 successive points on the same side of the center line;
- a cycle or pattern that repeats itself; and/or
- patterns that may appear which are unnatural.

Fig. Appendix 1-6 shows several such cases.

It is recommended to review the R chart first, as it is more sensitive to changes. If defective parts start appearing in a process, they will affect the R chart. The variation will increase, so some points will be higher than normal. The lower the points in the R chart, the more uniform the process. Two machines producing the same part may produce different forms of the R chart. Any changes in the process, such as operator inexperience, poor material, tool wear, or lack of maintenance will tend to shift points upwards.

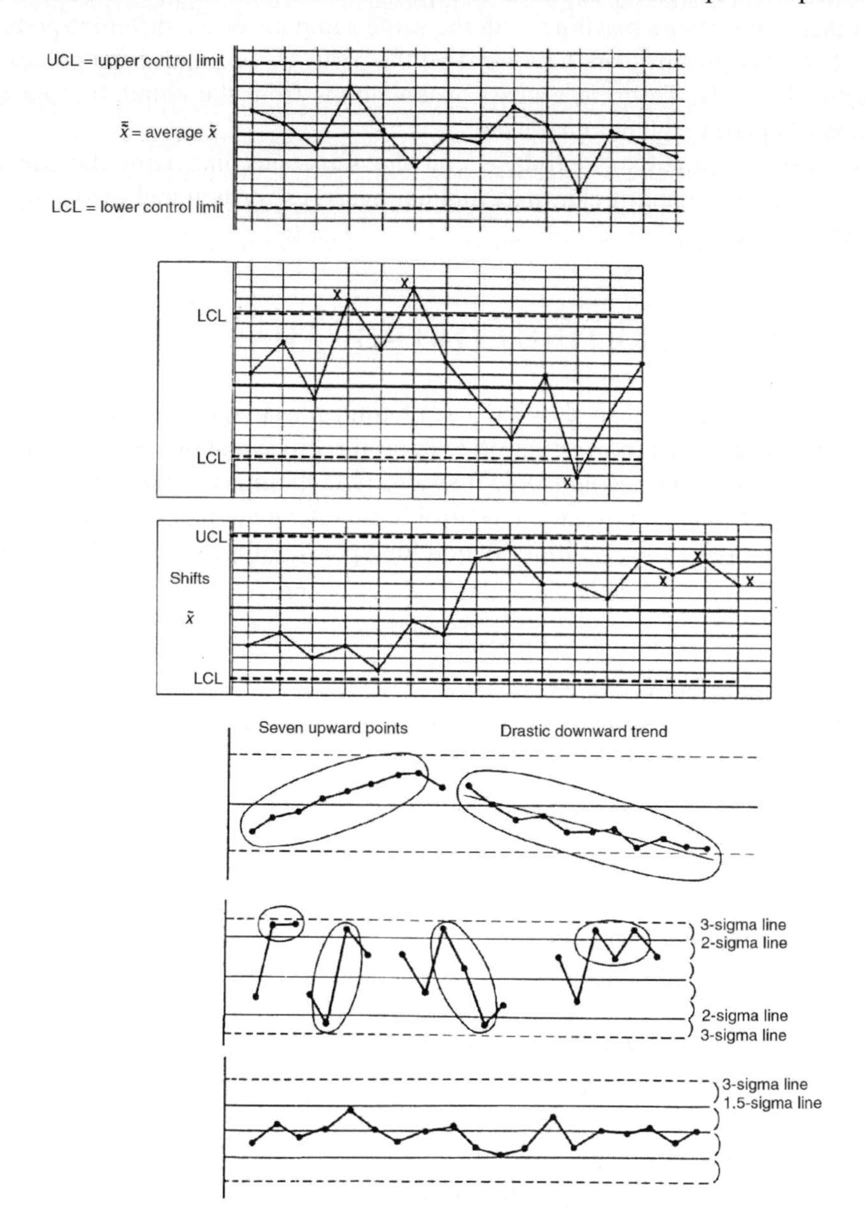

Figure Appendix – 1-6. Control chart interpretation

When the R chart is unstable, the $\bar{x}$ chart can be very misleading. With a stable R chart the variations on the $\bar{x}$ chart might be due to change of material, temperature change, new tool, machine setup, gradual tool wear, etc.

SPC controls the process and not the parts. Therefore any pattern on the charts that is not normal from a statistical point of view should be an alarm, i.e. something has changed in the process. Action must be taken to correct it. There is a 50% chance of having one point above the centerline and one point below it. There is also a chance that 68.25% of the readings will be around the centerline, and so on. Any deviation from such criteria should alert the user.

When a point is out of the control limits, the process should be stopped and immediate action should be taken to remove the cause, before more incorrect pieces are produced.

When there is a trend, the pieces are still within control, however in a short period of time they will run out of control. Therefore the process might keep on producing parts, but action should be taken to determine the cause and eliminate it.

When all points on the $\bar{x}$ chart are within $\pm\sigma$ and very low on the R chart, it means that all pieces have been produced within control. Statistically, however, the result is too good and is therefore impossible! The process may continue to produce pieces, but action should be taken to learn the cause of this shift in the process. A sticking gauge might be the cause, or a new material or a tool.

Action should be taken whenever the process does not behave according to statistical laws, even if it improves the process.

5. CAUSE AND EFFECT ANALYSIS - TROUBLESHOOTING

The statistics role in SPC is to spot variation in the process and to alert the operator to take action when needed. However, it does not tell what action to take. The action is based on technology and is handled by cause and effect study or troubleshooting technique.

The Pareto diagram is used to highlight the few most important causes and to highlight the results of improvements, and assists in determining the relative importance of cause. It can be the first useful document after data collection. Most of the defective items are usually due to a very small number of causes. Thus, if the causes of these vital few defects are identified, most losses can be eliminated by concentrating on these particular ones, leaving aside the other trivial defects for the time being.

Fig. Appendix 1-7 shows a Pareto diagram.

Once the problem is defined, a cause and effect analysis should be carried out in order to determine the actions to be taken to remedy the problem.

Type of Defect	Tally	Total
Crack	卌 卌	10
Scratch	卌 卌 卌 卌 .... 卌 //	42
Stain	卌 /	6
Strain	卌 卌 卌 卌 .... 卌 ////	104
Gap	////	4
Pinhole	卌 卌 卌 卌	20
Others	卌 卌 ////	14
Total		200

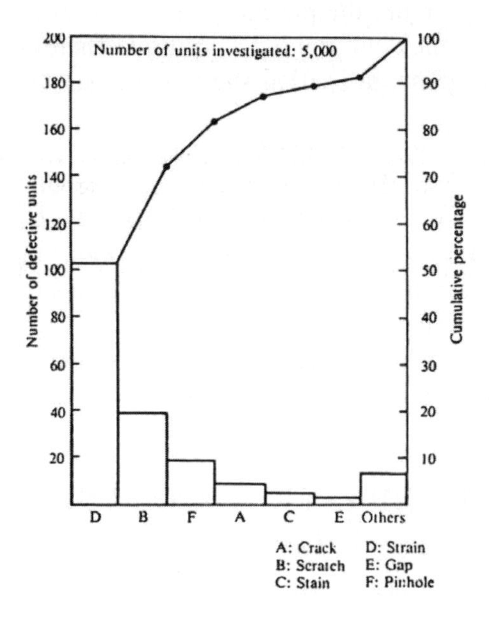

Figure Appendix – 1-7. Pareto diagram

The more widely used approaches to industrial troubleshooting are as followed:

- The 'what changed?' approach
- Conventional approach
- Checklist
- Kepner Tregoe approach
- Morphological approach
- Brainstorming

- Weighted-factor analysis
- Quality circle method (fishbone diagram)
- Relevance tree
- Statistical approach (experiment design)
- Simulation
- Expert systems.

Variations in products will always be present. The causes may be due to design, process, operator, or assignable causes. The last two are easy to analyze and remedy. Usually, the 'What Changed?' approach will give good results. This is used for products that have been manufactured for quite some time and with good results, but which have changed. If the effect is changed, then a cause must have occurred at the same time. A quick review of what has most recently changed often provides the clue to the underlying problem. Therefore it is recommended to conduct a system event log book at each workstation, as shown in Table Appendix 1-1.

Table Appendix – 1-1. System events log book

Work station No. Name:

Date	*Time/Shift*	*Changes: Oper/insp*	*Changes: Tools/Mat.l*	*Special notes*

*Special notes on any event such as:

- electricity failure	- tool breakage
- visitors interruption	- new measuring instrument
- accidents	- new set up.

The conventional approach is the one most people use to solve problems. Hypotheses concerning the cause are made, and some potential solutions are developed based on common knowledge. Trials are carried out until the solution is found. This approach generally is not a good approach for new employees or for personnel who are not familiar with the specific operation.

The other approaches are basically aimed at directing the problem solver to think in a systematic way.

Appendix - 2

PRODUCTION PLANNING - EXAMPLE

Stock allocation

1. INTRODUCTION

This appendix presents an example that demonstrates the flexible production planning method, based on PPTM as described in chapter 10. For simplicity and clarification two orders are introduced: Order #1; for 120 units of product 11 to be delivered on day 40; order #2 for 40 units of product B5 to be delivered on day 35. A brief description of the 15 resources which are used in this example is provided in Fig. Appendix 2-1.

Num.	Resource description	Power KW	Speed RPM	Handle time	Relative cost
1	Machining center	35	1500	1.10	4.0
2	Large CNC milling	35	1200	1.15	3.0
3	Manual milling	15	1500	1.66	1.4
4	Small drill press	2.5	1200	1.5	1.0
5	Old milling machine	15	2400	2.0	1.0
6	Small CNC milling	10	3000	1.25	2.0
7	CNC Lathe	25	3000	1.15	3.0
8	Manual new lathe	15	3000	1.42	2.0
9	Manual old lathe	10	3000	1.66	1.0
10	Circular saw				1.0
11	Band saw				1.0
12	Hack saw				1.0
13	Manual assembly				1.0
14	Machine assembly				1.5
15	Robotics assembly				3.0

Figure Appendix 2-1. The available Resources

Fig. Appendix 2-2 shows the bill-of-materials of these two products.

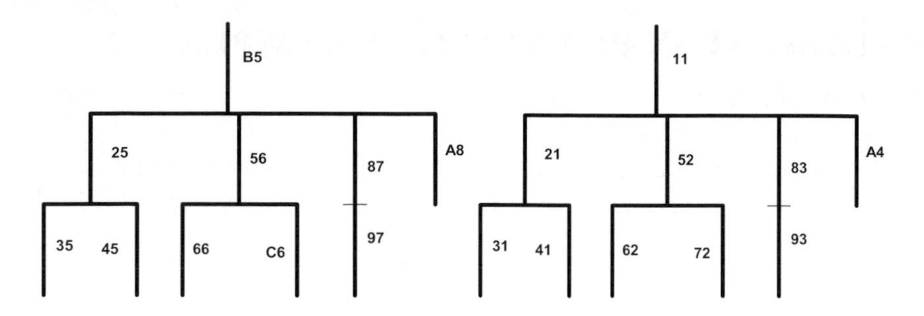

Figure Appendix 2-2. Bill-of-materials of Product 11 & B5

The first step objective is to set priorities for stock allocation. It works backward from the delivery date of each order. It starts with the assembly of the order, let's say order #1. The process planning table PPT is called to generate a process for the assembly of 120 units of product 11. The generated routing called for 1.31 days. The delivery date is day 40; therefore the assembly must start on day (40-1.31) 38.69. The next item on the bill-of-materials is item 21, the quantity is 120 units; the PPT is called to generate a process. The generated routing calls for 3.87 days. The item finish date is 38.69; therefore it must start on day (38.69-3.87) 34.82.

Similar computations are done for all orders and all items on the bill-of-materials. The results are detailed in the first column of Table Appendix 2-1; The Product Tree in Table format: Stock Allocation.

2. STOCK ALLOCATION

The objective of this step is to allocate available stock and to adjust the quantities of each item on the ordered products. The term "available stock" includes items in the storeroom, on the shop floor, in the purchasing process, in inspection and etc. The allocation is carried out with the product tree and gives priority to the critical path. The program scans all paths of the bill-of-material trees in a search for a low level item with the earliest starting day. In this example, as indicated in Table Appendix 2-1, (first step) item 41 of order #1 has a starting date of 29.67, which is the earliest one. Following the bill-of-material tree, from this item onwards, indicates that this item is part of product 11, order #1. Therefore, order #1 is the critical one and stock allocation will start with this order. It starts with the high level items and proceeds towards the low level items. Hence, initially item 11 will search for available stock.

In this example the available stock is as follows:

Item	11	41	52	62	83	93	A5	B5	C6
Quantity	40	10	18	9	11	20	28	20	15

Therefore, the quantity for item 11 will be reduced to 80 pieces, and the quantities of all items in the product tree will be adjusted accordingly. The PPTM will be called to generate a process plan for the new quantity, in the same manner as before. This is shown in second step in table appendix 2-1.

Table Appendix 2-1. The product tree in table format: stock allocation

			First step		Second step		Third step	
Order	Date	Item	Quant.	Start date	Quant.	Start date	Quant.	Start date
1	40		120					
		11	120	38.69	80	39.10	80	39.10
		21	120	34.82	80	36.55	80	36.55
		31	120	31.38	80	34.20	80	34.20
		41	120	**29.67**	80	33.04	80	**33.04**
		52	120	34.32	80	36.22	80	36.22
		62	120	31.50	80	34.28	80	34.28
		72	120	29.79	80	33.12	80	33.12
		83	120	37.64	80	38.37	80	38.37
		93	120	36.41	80	37.50	80	37.50
		A4	120	37.85	80	38.52	80	38.52
2	35		40					
		B5	40	34.52	40	34.52	20	34.73
		25	40	33.24	40	33.24	20	34.12
		35	40	31.97	40	31.97	20	33.56
		45	40	31.36	40	**31.36**	20	33.22
		56	40	33.07	40	33.07	20	34.03
		66	40	32.01	40	32.01	20	33.56
		C6	40	31.40	40	31.40	20	33.22
		87	40	34.09	40	34.09	20	34.52
		97	40	33.60	40	33.60	20	34.21
		A8	40	34.20	40	34.20	20	34.54

The program scans, again, all paths of the bill-of-material trees in a search for a low level item with the earliest starting day. In this example, as indicated in the second step in Table Appendix 2-1, the critical path is changed. The earliest starting date is now for item 45 of order #2 (day 31.36). Following the bill-of-material tree, from this item onwards, indicates that this item is part of product B5, order #2. Therefore, order #2 is the critical one and stock allocation will start with it. Hence, initially product B5 will search for available stock. There are 20 products in stock and they are allocated to order #2. Therefore, the quantity for item B5 will be reduced to

20 pieces, and the quantities of all items in the product tree will be adjusted accordingly. The PPTM will be called to generate a process plan for the new quantity, in the same manner as before. This is shown in the third step in Table Appendix 2-1.

The program scans, again, all paths of the bill-of-material trees in a search for a low level item with the earliest starting date. In this example, as indicated in the third step in Table Appendix 2-1, the critical path is changed. The earliest starting date is item 41 of order #1 (day 33.04). Following the bill-of-material tree, from this item onwards, indicates that this item is part of product 11, order #1. Therefore, order #1 is the critical one and stock allocation will start with it. Hence, initially product 11 will search for available stock. As item 11 was already allocated the next item of order #1 is item 21. As there is no item 21 in stock, the bill of material proceeds to item 31. As there is no stock for item 31, the bill of material proceeds to item 41. There are 10 units of item 41 in stock and they are allocated to item 4 of order #1. The quantity of this item is reduced by 10 to 70 required units. The PPTM is called to generate a process for 70 items 4 and the starting day is adjusted accordingly, as shown in the fourth step in Table Appendix 2-2.

Stock allocation proceeds to cover all orders and all items, the final state after stock allocation is shown in the final step in Table Appendix 2-2.

This final step is the working bill of materials of all orders. The next step is capacity planning.

3. CAPACITY PLANNING - RESOURCE LOADING

The working product structure lists all items that must be processed. It details the required quantity of each item and its relationship to other items.

The working product tree is not similar to the master product tree, as some items might be missing altogether, others might have a different quantity. The starting day on the list indicates the priority of machine loading. The item with the earliest starting day points to the order which is the critical order; therefore it should be loaded first. The loading is forward, from the low-level item, toward the first subassembly. Then all other items that are needed for assembly are loaded, and so on.

In our example, Table Appendix 2-2 final step, item 41 is the item with the earliest starting date (33.17). Therefore it will be the loaded first. The sequence of items to be loaded are 41, 31, 21, 72, 62, 52, 93, 83, A4, 11. Next order #2 is treated. The earliest start day is for item 45 (33.22); therefore the loading sequence will be items 45, 35, 25, 76, C6, 56, 97, 87, A8, B5.

Table Appendix 2-2. The product Tree in Table Format: Stock Allocation

Order	Date	Item	Fourth step Quant.	Fourth step Start date	Fifth step Quant.	Fifth step Start date	Final step Quant.	Final step Start date
1	40		120					
		11	80	39.10	80	39.10	80	39.10
		21	80	36.55	80	36.55	80	36.55
		31	80	34.10	80	34.20	80	34.20
		41	80	33.17	70	**33.17**	70	**33.17**
		52	80	36.22	62	36.03	62	36.93
		62	80	34.28	53	35.58	53	35.58
		72	80	**33.12**	53	34.79	53	34.79
		83	80	38.37	69	38.45	69	38.45
		93	80	37.50	49	37.87	49	37.87
		A4	80	38.52	52	38.70	52	38.70
2	35		40					
		B5	20	34.73	20	34.73	20	34.73
		25	20	34.12	20	34.12	20	34.12
		35	20	33.56	20	33.56	20	33.56
		45	20	33.22	20	33.22	20	33.22
		56	20	34.03	20	34.03	20	34.03
		66	20	33.56	20	33.56	5	33.87
		C6	20	33.22	20	33.22	5	33.74
		87	20	34.52	20	34.52	20	34.52
		97	20	34.21	20	34.21	20	34.21
		A8	20	34.54	20	34.54	20	34.54

The loading is recorded in a table of 15 columns, representing the 15 resources, and the rows represent periods of 120 minutes each. The entry in each area location indicates if the resource is idle or occupied. A blank indicates an idle period; otherwise the item names that occupy the resource at that period are listed.

Initially, the items that are in operation on shop floor are listed in the table. In this example it is assumed that there are no in-process items and the resource-period table is all blanks.

3.1 Resource loading for minimum cost process plan

The critical order is order #1 and the critical item is item 41. The quantity required is 70 units. The PPT is called to generate a process for 70 units of item 41. The recommended process plan is to use resource 11 for 7.03 minutes multiply by 70 units = 491.96 min. divided by period of 120 min. = 4 periods. As this operation can be made at any period, the systems scans resource 11 column and find that period 1 is available (empty) for 4 periods.

Therefore, the first 4 periods are allocated to perform item 41 as marked on Fig. appendix 2-3.

	1	2	3	4	5	6	7	8	9	10	11	12	13	14	15
1								A8	93		**41**				
2								93	97		**41**				
3								97	83		**41**				
4								A4			**41**				
5					**31**			A8			*72*				
6					**31**			87			*72*				
7					**31**						*72*				
8					**31**						45				
9					**31**						76				
10					**31**										
11			**31**		*62*										
12				**31**	*62*										
13				**31**	*62*										
14			*62*		**21**										
15			C6	*62*	**21**										
16			56		**21**										
17			56		**21**										
18			56		**21**										
19					**21**								B5		
20					**21**										
21			**21**		*52*										
22			**21**		*52*										
23			**21**		*52*										
24			52												
25			52												
26			52												
27			52												
28			52												
29													11		
30													11		
31													11		
32													11		
33															
34															
35															

Total relative cost = 76.2

Figure Appendix 2-3. Load profile for minimum cost process planning

Next item 31 is treated. The quantity is 80 units. The PPT is called and a minimum cost process plan is generated, it calls for three operations on:

Resource 5 for 8.915 min. x 80 / 120 = 5.94 periods
Resource 3 for 1.83 min. x 80 / 120 = 1.22 periods
Resource 4 for 2.995 min. x 80 / 120 = 2.00 periods

The loading starts with Resource 5. The loading can start only after item 41 is finished (i.e. period 5). As Resource 5 is available at those periods it is loaded at period 5 to 10 and marked as 31. Resource 3; the loading can start only after resource 5 finishes (i.e. period 11). Therefore period 11 of resource 3 is marked by 31. Resource 4; the loading can start only after resource 3 finishes (i.e. period 12). Therefore period 12 &13 of resource 4 are marked by 31.

Next item 21 is treated. The quantity is 80 units. The PPT is called and a minimum cost process plan is generated, it calls for two operations on:

Resource 5 for 10.83 min. x 80 / 120 = 7.2 periods
Resource 3 for 4.50 min. x 80 / 120 = 3.00 periods

The loading may start only after item 31 was finished, i.e. period 14. As machine 5 is available at those periods it is loaded at period 14 to 20 and marked as 21. Next resource 3 is loaded after resource 5 finishes (i.e. period 20). Therefore periods 21, 22, 23 on resource 3 are loaded and marked by 21.

Next item 72 is treated. The quantity is 53 units; the PPT is called and generates a process plan on resource 11 that requires 3 periods. The operation can start at period 1, however, resource 11 is occupied at periods 1 to 4, (these periods are marked by 41) and the earliest free period is period 5. Therefore item 72 is loaded at periods 5 to 7 on resource 11 and marked 72.

The same loading procedure proceeds till all items are loaded. Fig. appendix 2-3 shows the loading of the two orders, by using process planning minimum cost criterion of optimization.

3.2 Resource loading for maximum production process plan

In this case, the PPT is called to generate a routing using the maximum production criterion of process planning, instead of the minimum cost criterion of optimization.

The loading method is the same as before. The load profile is presented in Fig. appendix 2-4. Examining this figure shows that the shortest processing time of all milling operations are when resource 1 is used. It is clear that the best resource for each item is selected.

	1	2	3	4	5	6	7	8	9	10	11	12	13	14	15
1							93				41				
2							93				**41**				
3							83				**41**				
4							83				**41**				
5	31						A4				*72*				
6	31						97				*72*				
7	31						97				*72*				
8	31						A8				45				
9	31										76				
10	21														
11	21														
12	21														
13	21														
14	21														
15	21														
16	21														
17	21														
18	62														
19	62														
20	62														
21	62														
22	52														
23	52														
24	52														
25	52														
26	52														
27	52														
28	35														11
29	35														11
30	25														
31	24														
32	C6														
33	56														
34	56														
35															B5

Total relative cost = 162

Figure Appendix 2-4. Load profile for maximum production process plan

It is strange that when working with the maximum production criterion of optimization, the lead time to manufacture the product mix, (35 periods) is longer than the lead time when using the minimum cost criterion of optimization (32 periods). This phenomenon is due to the fact that in the maximum production criterion of optimization all items are planned to be processed on the best resource, (i.e. resource 1); as a result, a long queue of work piles up. Maximum production refers to processing of any individual item, but not a product mix. Optimization of product mix is a task of production planning and not that of process planning.

3.3 Resource loading - flexible methods

A simple case of process interchange is when the criterion of optimization interchanges: first minimum cost is used and the next branch of the product structure uses the maximum production criterion of optimization to generate a process plan by the PPT. The lead time to produce the required product mix in this case is 23 periods. The load profile in this case is shown in Fig. appendix 2-5.

A more sophisticated capacity loading algorithm is to select a different process plan whenever an item has to wait for a resource. Each plant may use any economic algorithm he desires. It may be dependent on the number of periods that an item must wait for an idle machine, or the cost differences or a combination of the two.

In this example the objective is to produce the product mix of the two orders in the shortest lead time possible while keeping the processing cost to a minimum (as a second objective). The rule used by the algorithm is to change a process whenever an item has to wait, even for one period. The least cost resource is selected if it does not affect the lead time.

For example: item 41 is on the critical path. Therefore it is loaded on machine 11 for 4 periods. To meet the least lead time, the system used the maximum production criterion of optimization to generate the process plan for item 31. It calls for resource 1 for six periods, and it may start only after item 41 is done (i.e. period 5). As resource 1 is idle at that period, it was loaded with item 31 and marked 3. A check is made to determine if resource 2 or 6 or 3 can process this item in the same number of periods. The check reveals that producing item 31 on resource 2, will take seven periods, on resource 6 will take seven periods, and on resource 3 will take eight periods. Therefore, item 31 is loaded on resource 1.

Next item 21 is considered, again with the maximum production criterion of optimization. The PPT recommends using resource 1 for seven periods. It can start only after item 31 is done (i.e. period 11). Resource 1 is idle at that period. However, resource 2 can also process this item also in seven periods (note that the number of periods is rounded to the closer integer). The hourly rate of resource 2 is 75% of that of resource 1; as there is no saving in lead time by using resource 1, it is decided to use resource 2 for item 21 and the periods 11 to 17 are marked by 21.

	1	2	3	4	5	6	7	8	9	10	11	12	13	14	15
1							93	A4			41				
2							93	A4			41				
3							83	A8			41				
4							83				41				
5	31						97				**72**				
6	31						87				**72**				
7	31										**72**				
8	31				**62**						45				
9	31				**62**						76				
10	21		C6		**62**										
11	21		**62**												
12	21		56	**62**											
13	21		56		**52**										
14	21		*56*		**52**										
15	21				**52**										
16	21		**52**												
17	21		**52**												
18	35		**52**												
19	35		**52**												
20	25		**52**												
21	25														11
22															11
23															B5
24															
25															
26															
27															
28															
29															
30															
31															
32															
33															
34															
35															

Total relative cost = 131

Figure Appendix 2-5. Load profile for minimum cost/maximum production process plan

Next, item 72 is considered for loading. The PPT recommends using resource 11 for three periods; it can start at period 1. It scans resource 11 and finds that the resource will be idle only in period 5. Therefore it turns to the PPT and generates another process without using resource 11. This time the PPT recommended using resource 10. The processing time increases, but the processing can start without delay in period 1. It will take four periods on resource 10, instead of three periods on resource 11. By making this change item 72 can be finished three periods earlier.

The next item to be loaded is item 62. The PPT recommends using resource 1 for four periods. It can start only after item 72 is done (i.e. period 5). However resource 1 is occupied till period 10. Therefore, instead of waiting for six periods (5 to 10) the PPT blocks resource 1 and generate another process which recommends using resource 2 for four periods. Resource 2 is available and period 5 to 8 are marked as 62.

The next item to be loaded is item 52. The PPT recommends using resource 1 for five periods. It can start only after item 62 is done (i.e. period 9). However resource 1 is occupied till period 10. Therefore, instead of waiting for 2 periods (9 and10) the PPT blocks resource 1 and generates another process which recommends using resource 2 for five periods starting at period 9. However resource 2 is available for only two periods, and is occupied in periods 11 to 17. Therefore instead of waiting for 9 periods (9 to 17) the PPT blocks resource 2 as well and generates another process which recommends using resource 6 for six periods starting after item 62 is done (i.e. period 9). Resource 6 is idle at those periods and item 52 is loaded, and periods 9 to 15 are marked as 52.

The assembly of product 11 may start only after all components are available. Item 21 is the critical one and is done at period 17, which means that the assembly of item 11 can start only in period 18 and calls for three periods. Item 52 is ready for assembly in period 15, which means that it waits for three periods before it can be used for the assembly. Therefore, the system checks if the cost of processing can be reduced without causing any increase in the lead time. Item 72 is already moved to the first period. Therefore item 6 is being evaluated. The PPT is called to generate a process plan with minimum cost criterion of optimization for item 62. The recommended process calls for using:

Resource 5 for three periods
Resource 3 for one period
Resource 4 for one period

It takes a total of five periods, which means increase of one period and reduce the cost from 12 units to 5.3 units Therefore this new recommendation is to move item 62 from resource 2 to resources 5, 3, and 4. Period 5, 6 , and 7 of resource 5 are marked 62; period 8 of resource 3 is marked 62; and period 9 of resource 4 is marked 62.

Similar treatment to item 52 indicates that it may use a minimum cost optimization process which calls, as computed before for three periods of resource 5 and five periods of resource 3. This reduces the cost from using resource 6 from 14 units to 10 units. It is economical without affecting the lead time of the assembly, and the change is being made (i.e. the load is shifted from resource 6 for periods 9 to 15, and periods 10 to 12 of resource 5 are marked 52 and periods 13 to 17 of resource 3 are marked 52.

The system treats all other items with a similar logic and the results are shown in Fig. appendix 2-6.

	1	2	3	4	5	6	7	8	9	10	11	12	13	14	15
1								A8	93	*72*	**41**				
2								93	97	*72*	**41**				
3								97	83	*72*	**41**				
4								A4		*72*	**41**				
5	**31**				***62***			A4		76	*45*				
6	**31**		35		***62***	C6		87							
7	**31**		35		***62***	56									
8	**31**		*62*			56									
9	**31**		25	62											
10	**31**		25		***52***										
11		**21**			*52*								B5		
12		**21**			*52*										
13		**21**	*52*												
14		**21**	*52*												
15		**21**	*52*												
16		**21**	*52*												
17		**21**	*52*												
18													11		
19													11		
20													11		
21													11		
22															
23															
24															
25															
26															
27															
28															
29															
30															

Total relative cost =101

Figure Appendix 2-6. Load profile for case of variable process plan

Carrying out the planning actions as described above, results in:

- Minimum processing lead time
- Meeting delivery date
- Resource utilization
- Minimum work in process
- Minimum capital tied down in production

Eliminate bottlenecks in production.

Index